Science Education from People for People

Contributing to the social justice agenda of redefining what science is and what it means in the everyday lives of people, this book

- introduces science educators to various dimensions of viewing science and scientific literacy from the standpoint of the learner, engaged with real everyday concerns within or outside school;
- develops a new form of scholarship based on the dialogic nature of science as process and product; and
- achieves these two objectives in a readable but scholarly way.

The authors want science education to be *for* people rather than strictly *about* how knowledge gets into their heads. Opposing the tendency to teach and do research as if science, science education, and scientific literacy could be imposed from the outside, they discuss applications of epistemologies not often recognized in science education, and offer an opposite position to the rhetoric of "No Child Left Behind" and its top-down approach to mandating what students need to know. Taking up the challenges of this orientation, science educators can begin to make inroads into the currently widespread irrelevance of science in the everyday lives of people.

Designed as a forum in which leading scholars present and interact about issues arising from the concept of scholarship from people for people, utmost attention has been given to making this book readable by the people from whose lives the topics of the chapters emerge, all the while retaining academic integrity and high-level scholarship.

Wolff-Michael Roth is Lansdowne Professor of Applied Cognitive Science at the University of Victoria, Canada.

Contents

Illustrations

Table

Figures

Contributors

Angela Calabrese Barton is Associate Professor of Science Education at Michigan State University. Her research in urban science education has focused on high poverty urban middle school youths' scientific literacies in and out of school, and on the preparation of teachers to teach science in high poverty urban communities.

Carol B. Brandt is Assistant Professor of Social Foundations at Virginia Polytechnic Institute and State University, Blacksburg. Her research and teaching examines the interplay among language, local knowledge, and science in contexts beyond the classroom (farm fields, kitchens, molecular biology laboratories, science museums, and parks).

Katherine Richardson Bruna is Assistant Professor of Multicultural and International Curriculum Studies at Iowa State University. A former bilingual educator and evaluator of educational policy and practice related to bilingual students in California, she is interested in issues related to culture, language, identity, and equity in schooling.

Christopher Emdin is Associate Professor of Science Education at Teachers College, Columbia University. He conducts research on issues of race, class, and diversity in urban science and mathematics classrooms. His most recent work explores new theoretical approaches to research in urban science education and student/teacher identity formation around science.

SungWon Hwang is Assistant Professor of Science Education at the National Institute of Education, Singapore. Prior to that she was postdoctoral fellow at the University of Victoria, where she conducted interdisciplinary research projects that articulate dialectic frameworks of learning and identity in the context of science and technological work.

Hannah Lewis, MS, is Assistant Scientist in Sociology and Sustainable Agriculture at Iowa State University. She coordinates bilingual programming to develop small-farm entrepreneurship among immigrants in diversified community-based food systems.

Eileen Carlton Parsons is Assistant Professor of Science Education at the University of North Carolina–Chapel Hill. She conducts research on the teaching and learning of science in the early and middle grades with respect to context, culture, and race.

Wolff-Michael Roth is Lansdowne Professor of Applied Cognitive Science at the University of Victoria. He conducts research on learning science and mathematics across the lifespan, from kindergarten to professional praxis (including science laboratories, fish hatcheries) from multiple theoretical perspectives (phenomenology, sociology, cultural studies, cognitive science) and for multiple communities of practice, including science studies, sociology, linguistics (pragmatics), learning sciences, and education.

Karen L. Tonso is Associate Professor of Social Foundations at Wayne State University (Detroit), studies engineering education by melding fifteen years in engineering with degrees in mathematics education, and anthropology of education. Her research examines relations of power, related to gender, race/ethnicity, social class, or other forms of life.

Preface

This book has three major objectives: (a) to introduce science educators to the various dimensions of viewing science and scientific literacy from the standpoint of real people; (b) to develop a new form of scholarship that is based on the dialogic nature of science as process and product for real, everyday people; and (c) to achieve the two previous objectives in a readable but scholarly way. In the same way as the sociologist Dorothy E. Smith, who takes a women's standpoint to her research, the authors of this volume take a standpoint that begins in the actualities of people's lives, their own or those of others with whom they are working closely. The results of these investigations are intended to be useful to those very people, not via a detour through academic discourse, that is showered upon and re-introduced to the people's lives by means of an "application" or "implication," but by means of a discourse that never leaves the lives of real people. As a result, the authors arrive at framing science and scientific literacy, and therefore also science education, in terms of everyday people, who become the sites of consciousness, mind, participative thinking, subjectivity, agency, identity, and so forth as a result of their doings. All contributors have experience in "writing from the margins"—to make the marginal position central in their perspective on science education. They are therefore well-positioned to write a science education *from* the people designed to be *for* the people rather than conforming to some *external* standard that pays lip service to taking into account the lives and experiences of the learner.

From the beginning, I planned this book as both very readable and very articulate about all matters of identity concerning science, science education, and science learning. I wanted a book that is grounded in the everyday experiences of different people in different parts of the continent and from different cultural backgrounds. During its conception, I was thinking about a book that is not simply a collection of a number of chapters that look more like journal articles with little connection between them. My vision was more like a forum, in which leading scholars present and interact about issues arising from the concept of scholarship from people for people.

To achieve this goal—and consistent with the proposed subtitle of this book—the authors, especially in the two *metalogue* sections, actively think through and propose alternative approaches to science and science education *for* those people with educational agendas: teachers, parents, informal educators, non-governmental organizations, and others. This, then, is a third mode of subjectivity and activity, for the authors engage in making available resources and giving directions for *changing* their own and others' lifeworld rather than simply seeking to understand it (as in traditional quantitative and qualitative research).

This book is designed to be useful not only to a small group of initiates, people with the great level of expertise of the contributors, not only to a restricted wider scholarly audience, not only to graduate students (who are novices in the field of educational research), not only to colleagues not specializing in questions of research methods and methodology, and to academically trained policy makers including those who work in funding agencies. Rather, the book is intended for an intelligent and informed readership generally. All contributors' utmost attention has been given to producing their individual chapters and the book as a whole to be readable *by the people in whose lives the topics of our chapters emerge*, all the while retaining academic integrity and high-level scholarship.

Ways in which the authors achieve greater readability on the part of a broader audience is by (a) a reduced number of references in the text—additional references should be requested directly from the authors; and (b) the use of a language that non-specialists recognize as their own and a build-up (explication) of those important concepts that the chapter texts intend to convey—again, the text was held at a more general level and if needed, more specialist terminology, technical vocabulary, and explication are placed in footnotes.

Wolff-Michael Roth
Victoria, British Columbia
August 2008

1 Taking a Stand(point)

Introduction to a Science (Education) from People for People

Wolff-Michael Roth

The past 50 years have seen tremendous activity in science education, both in terms of the development of curricula and in terms of the research conducted on how people (mostly school students) know and learn science. Yet despite the tremendous amount of work done, many of the problems that had occasioned interest in the field after the Sputnik shock continue to persist. Thus, more than ever, many students do not see science as relevant to their lives and opt not to enroll in science courses at the secondary and post-secondary levels. More than ever, students do not opt for careers in science and scientists, reflecting on this issue in their flagship journal *Science*, wonder about ways in which they can increase the "throughput" in their science "pipelines." A concern for science education that is to serve and educate *all* members of society, however, cannot be the same as the one for throughput and pipelines. Whereas scientists' concerns are legitimate to the extent that we need scientists and engineers to produce knowledge that allows humans to control their environment and therefore to guarantee the survival of the species, science education *for all* has to be different in nature (Roth & Barton, 2004) because it has to address itself to the very different needs that distinguish the general public from those specific individuals whose needs are met when they pick up careers in science.

In this book, the authors as a collective therefore are not concerned with *throughput* or *filling the pipeline.* Rather, collectively they take the position that a key problem of past efforts is this: educators, psychologists, and natural scientists defined the nature of science, science education, and scientific literacy in terms of the products of laboratory science. The definitions of science have always been in terms of science content from a scientific perspective and in terms of disembodied forms of knowing. The definitions had little or anything to say about the tremendous experiences and competence everyday people (including students) have especially when they are uninstructed in science; and they had little to say about how science and science education could assist everyday, ordinary, and just plain folk in and with the problematic situations that they face in their ongoing lives. There are many such problems, as shown for example in the issues of (Zuni) gardening, having children, or facing (chronic) illness that feature in some of the chapters of this book. Yet these problems rarely if ever demand the kinds of knowledge that students are to acquire in their science classes. Those who cope with illness do not need to know the Krebs cycle or Newton's third law; nor do Zuni gardeners need to know this form of science, as their own ways of gardening corn already is so much more adapted than the scientifically informed ways of industrial farmers. In the chapters of this book, therefore, there is little about how to cram—by transfer or construction—atoms, molecules, Krebs cycle, and Newton's third law into the heads of children. And these problems always are bound up with human beings, lived experiences, emotions, worries, effect–affect transactions, and so forth.

One of the questions some science educators concerned with science education and social justice ask is how to make the sciences more relevant to students specifically, and all members of society more generally. But how do we have to think, and think about science, so that it becomes more relevant? Certainly not in the same ways that have turned students away from the sciences for the past five decades since Sputnik was launched. With a re-orientation of science and scientific literacy in and through problematic issues in the lives of people, science educators might actually begin to make inroads into the currently intractable problem of the irrelevance of science in the everyday lives of students specifically and all everyday folks more generally. Science would be relevant in and to these lives if the people themselves recognized it as a resource for action and therefore as something that *expands their room to maneuver* and power to act—i.e., to their agency. This concern for science as a useful resource in and for the lives of everyday people is at the heart of this book. That is, science education in the way the contributors approach it here is centrally about social justice rather than the stuffing of science content into the heads of children, students, and everyday folks. But the sciences have to be more. In a democratic society, the sciences have to be open to critique, open to be contested, unless they want to be of the same status as religions that one has to take on faith.

This tension for science educators arises from the fact that they understand their task as one of teaching canonical science. A quick look at the news shows, however, that science is not just a resource in everyday life but also a contested terrain. This is immediately evident when we follow the debate about global warming, where each side finds scientists to support their ontological stance according to which global warming exists (as the former vice president A. Gore suggests in his documentary *An Inconvenient Truth*) or does not exist (as G. W. Bush upheld for a long period of time). It is clear that the science itself is becoming the terrain that is contested in a debate (or "battle") where science also is rallied in support of the various and divergent arguments. Allowing students specifically, and all people more generally, to draw on their knowledges as a resource to contest other forms of knowledge in decision-making processes leads to further tensions because of the incommensurability of the knowledgeabilities involved. Thus, scientists, policy makers, politicians, and everyday folk find themselves struggling with "integrating" forms of knowledge that cannot be integrated because they cannot be reduced to one another (e.g., Roth, 2008b).

Science as Resource and Contested Terrrain

The purpose of this book is to oppose the general tendency of doing science education (teaching, research) as if science, science education, and scientific literacy could be imposed from the outside, and as if the pertinent forms of knowing and learning were independent of the human orientation toward expansion of their room to maneuver. That is, the contributors take the position that adult, adolescent, and child learners will find science, scientific literacy, and science education relevant once they see and understand how their own possibilities of acting and being in the world expand. Such expansion comes with a positive (emotional) valence, which is therefore an important mediating aspect of becoming and engaging as a (science) learner. Therefore, in this book the contributors aim at constructing perspectives on science, scientific literacy, and science education grounded *in the lives of real people* and that are oriented toward *being for real people* (rather than disembodied minds). Our concerns thereby intersect with those of Dorothy E. Smith (2005), who, in writing *Institutional Ethnography*, produced a sociology *for* people. Collectively, the authors in this volume want science education to be *for* people rather than *about* how knowledge gets into the heads of people—be it by means of construction, transfer, or internalization.

One proposal in the past has been that science itself has to become a contested terrain and resource (Roth & Barton, 2004). Taking such an approach no longer allows science educators to think about the ways in which we can fit students specifically, and the general public more generally, to science as it is practiced in laboratories and in scientific journals. This is a form of science that, despite its origin in the everyday pursuit, languages, and practices of people, has become a form of practice that elevates and imposes itself as something special. Scientists have become the new high priests in a secular society. Whatever they have evolved as practice is taken and presented as something like a gold standard against which all other practices are evaluated—the discourses of students *mis*conceptions, *alternative* frameworks, or *naïve* conceptions constitute ample proof for the deficit discourse science educators employ with respect to everyday knowing. For scholars in the cultural studies, of course, this is but another culturally specific standpoint on knowledge and on knowledge production and evolution. It does not have to be that way, as the events surrounding the AIDS community have shown, and how AIDS activists have been able to bring about a change in the methods of testing new drugs (e.g., Epstein, 1995).

The analysis of AIDS research has become an important testing ground for the social sciences as AIDS activism exerted a politics of identity organized by constituencies around specific illnesses and diseases such as breast cancer, chronic fatigue, and environmental illness. The relations between AIDS activists and scientists in particular showed how science itself can become both a resource (e.g., in the development of new drugs) and a contested ground (e.g., as the standard ways of doing science come to be scrutinized, questioned, and changed). Here, the AIDS activists worked from a particular position, that of people affected by the disease, and from the associated standpoint, communicated their point of view with such vigor that they were able to change how science is done and therefore what science is.

The standard method for testing the effectiveness of new drugs has been the double-blind experiment, randomization of participants to treatment and control, and working with particular populations. For example, AIDS trials employed samples consisting largely of middle-class white men. AIDS activists were able to argue that subject populations should be extended to injection drug users and hemophiliacs, women, minorities, and differing sexualities. They simultaneously pushed for (a) fair access to experimental drugs rather than random assignment and (b) generalizability. However, treatment activists have been able to engage scientists over the processes of drug testing and in the process have become legitimate players. Their legitimacy can be gauged from the fact that AIDS treatment activists have become, following a long struggle, full members of various committees of the U.S. National Institutes of Health that oversee drug development. They have also become participants in the advisory committee meetings of the U.S. Food and Drug Administration, where any new drug is considered for approval prior to being released.

The AIDS case is but one of a number of forms of activism that has had mediating influences on science and how it operates—i.e., its methods—and therefore on the very definition of science. Thus, organizations of people with a variety of diseases and illnesses have been able to assert their needs and mediate *what* science is done and *how* it is done: Those struck with illness *do have* power, as medical sociologists have shown (Rabeharisoa & Callon, 1999). Environmental activists have inserted themselves into the public debate and policy making concerning the testing and use of genetically modified organisms. And individuals from First Nations and just plain folks (e.g., fishermen in Newfoundland) around the world have begun to work with scientists and thereby brought about changes in the ways in which relevant systems are modeled, tested, and theorized. Thus, for example, the "Back to the Future" (e.g., Pauly, Pitcher, & Preikshot, 1998) approach uses complex computer-aided tools to combine vastly different forms of knowledge, such as the ones scientists produce in their laboratories and the local knowledge of Aboriginals and residents.

In each chapter that follows, the respective author/s take up the challenge of writing an approach to science, scientific literacy, and science education with a problem *relevant to one or more real persons* and to develop theory and a description of their approach out of this problem. This therefore becomes a science education *from the standpoint* of the knower/learner, engaged with real everyday concerns either within or outside school. That is, rather than developing a theoretical framework that will be imposed on some data materials, the authors begin with a problem in the lives of people (children, adolescents, adults) and then engage in a form of institutional ethnography, which begins with everyday experience as the grounds from which discoveries can be made. The resulting (and necessary) standpoint will be that of children, women, persons of color, Aboriginals, expecting mothers, or a person afflicted with chronic but undiagnosed illness. The authors thereby come to think and theorize science education from the place of those who learn and from the place of the people who might become interested because the payoffs from engaging with science in one or another form include an increase in their agential room to maneuver.

In accordance with Dorothy E. Smith's work, the important dimension of doing research from a particular "standpoint" is that it does not subordinate the knowing subject to forms of knowledge that have been objectified and codified into science textbooks, that is, to the societal-hierarchical forces in a political economy. The present authors allow us to think ethnographically (sociologically, anthropologically) from the place of real people (including themselves) struggling with one or another facet of daily life (including school life). Yet, as those in movements of previously (and present-day) marginalized groups know, there are experiences that discourse does not articulate, and institutional ethnography is one of the tools that can uncover and make thematic these experiences. This also requires social scientists to go beyond what is apparent to real people: like the concepts (ideologies) and artifacts that we have come to use, there are things in our lives that have a determinate effect on what we do. These concepts and artifacts have an insidious effect in the sense that they may go against the interests of real people, instead serving those in power and the ruling relations. For example, in one town of British Columbia about 15% of the students were on Ritalin because they were said to have ADHD (attention deficit hyperactive disorder). Surely, there is not 15% of a population afflicted; and there are other ways to deal with attention than drugging children. Here concept of ADHD appears to be used to subjugate and drug children, who are calmed to the point that they are submissive, and it has little to do with the real lives of these children. But parents are made to buy into the use of Ritalin simply because it is said to deal effectively with ADHD. This formula (name) therefore serves as a (discursive) tool to make parents and children buy into and therefore produce and reproduce a practice that ultimately only serves the pharmaceutical industry.

The chapter contributions in and to this book strive to bring together two modes of subjectivity and activity that ordinarily have been kept separate: our personal lives as mothers, ill persons, student in a science class, Aboriginals in science, environmentalists, etc. that we share with others and our professional lives as academics. In these latter lives, we have all too often tended to objectify knowledge (discourse), on the one hand, and those who know and learn, on the other hand. Consciousness thereby came to be stripped of local particularities, auto/biographies, contingencies, needs, and emotions of people to whom some form of science and scientific literacy could become a resource. (Here, we do not pre-specify the nature of science and scientific literacy but rather leave it open to rearticulate what their nature is as an outcome or implication of the work reported.) Especially in the two *metalogue* chapters, the authors actively think through and propose alternative approaches to science and science education *for* the people.

Content and Structure of this Book

The book consists of 12 chapters, including two discussion forums ("metalogues") and is grouped into two parts, "Culturing Knowledges" and "Othering the Self, Selfing the Other." In the following two subsections, I briefly describe, contextualize, and relate the contents of the chapters in the two parts.

Part I: Culturing Knowledges

The title of the first part of this book evokes the inseparable connection all knowledge has to culture with the *double entendre* that knowledges also have to be cultured (nurtured). Etymologically, the term "culture" derives from the participle of the Latin *colĕre*, to attend to, respect. The word then made it from the term *cultura*, cultivation, tending, and worship through the French culture (*couture*) into English. In its present-day use, besides being a theoretical term in cultural studies, anthropology, and sociology, the term also refers to the action of cultivating soil, tillage, rearing plants and animals. In an interesting article about culture and identity, the etymology and these other senses of the word are brought into play to argue against culture as something pure:

> "Cultures," or whatever we call by this name, do not add up. They encounter one another, mix with one another, alter one another, reconfigure one another. They cultivate one another, clear one another's ground, irrigate or drain one another, work one another or mutually graft themselves onto the other. (Nancy, 1993, p. 13, my translation)

As a result, there is nothing like *a* culture, because every entity thus denoted is itself multicultural and the result of a continual *mêlée, that is, of a* process "of affronting, confronting, transforming, detouring, developing, recomposing, combining, and doing bricolage" (p. 13). With culture, all of its elements are subject to the same processes, so that we cannot think of language or identity as self-same concepts denoting self-same phenomena. Identity, language, knowledge, and so on are heterogeneous *processes*, continuously making and remaking themselves, never quite themselves and always already other than themselves at the very instant that they realize one of their possibilities—in actualized identity, realized utterance or written sentence, concretely articulated and enacted knowledge. Historical developments of culture and language cannot be understood unless every (speech) act already is considered a change in and of what has been available up to the moment of its beginning (Bakhtin/Vološinov, 1973).

All chapters in this first part focus on the experience of science and on scientific knowledge at and across the border of different inherently heterogeneous cultures—African Americans (people of color) in a largely white society (Parsons, Emdin), Zuni Indians and Latino/as in the US, Asians (Korea, Japan) in Canada (Hwang, her participant). The chapters show that knowledges are not impersonal but fundamentally situated in and mediated by culture and language, themselves not unities or unicities but multiplexes and pluralities. Learning science therefore is more than appropriating a new code, and requires a reconfiguration of the Self or a reconfiguration of science. In any event, it requires a continuous hybridization of cultures and cultural knowledges.

In "Revisiting and Reconsidering Authenticity in Science Education: Theory and the Lived Experiences of Two African American Females," Eileen Carlton Parsons, an African American scholar, addresses the perceived universality of Western science and the way the promulgation of this universality in the practices of science education and scientific literacy serves to exclude many voices, ways of knowing, and ways of being that could potentially enhance science and its meaningfulness. In this chapter, Eileen literally places two narratives side by side. On the one side, there are the stories of two African American women brought up in two historically distinctive eras—an 81-year-old from the rural poor with an elementary education and a 30-year-old from the rural middle class who holds a PhD. On the other side, Eileen articulates a framework synthesized from two conceptual models in psychology, Western science as particularistic rather than universal. The juxtaposition of the two forms of text allows her to articulate and examine insights pertaining to life-based authenticity as it relates to Western science, science education, and scientific literacy.

In her chapter "Faith in a Seed: Social Memory, Local Knowledge, and Scientific Practice," Carol B. Brandt, who grew up in an agricultural community where growing one's own food was second nature in everyday life, is concerned with sites for learning science outside of schools that are framed and shaped by social, economic, and political discourses. Gardening was woven into the "common sense" of her community and, along with other youth, she attended 4-H meetings[1] organized by the county agricultural extension agents, and began to record "scientific data" on yields and to experiment with new varieties. Agriculture in this German-American community was infused with Christian values and a staunch faith in Eurocentric science as a result of the Progressive Era that revolutionized Midwestern farming in the 1920s. Attached to this local knowledge and informal science were discursive constructions of time and space that also ordered community practices. None of her experiences, however, prepared Carol for gardening in west central New Mexico. On the high desert plateau, the arid environment is marginal for growing conventional crops, and yet Zuni is the home to one of the oldest Indigenous agricultural traditions in the United States. When Carol began working with Zuni farmers and gardeners her notions of "common sense" in agricultural practices shifted while she examined gardening as the relationship among local knowledge, economics, and the political history of Eurocentric science in this Indigenous community. In her chapter, Carol chronicles how she grasped the role of social memory in maintaining local knowledge in an agricultural repertoire despite dramatic economic and social change. In learning how to garden in a Zuni way, she came to understand how Eurocentric science is part of a larger custodial discourse between the federal government and Indigenous people. Drawing from interviews with 50 gardeners and farmers at Zuni, she describes ways in which local knowledge is parsed and dispersed in the community, and how local conceptions of space-time is often at odds with Eurocentric science.

The mediations of learning that come with migrating from one country and culture to another, a prevalent experience in an increasingly globalized world, is the topic of chapter 4 entitled "Language and Experience of Self in Science and Transnational Migration." The two authors and their research participant have migrated to Canada, two from Asian countries (SungWon Hwang [Korea], Miko [Japan]) and one from Europe (Wolff-Michael Roth [Germany]), and all three have begun their lives in a culture other than the Anglo-Saxon Canadian that historically defined the area in which they now live—though in many homes today English is not the language of choice (in Vancouver, less than 45% of families speak English at home). Moving from one culture to another is not a problem from rationalist perspectives, because it simply involves changing from one system of codes (language, culture) into another. From such a perspective, all one has to learn is how to translate between the two forms of code. All three individuals in this chapter have experienced how transnational migrations both within Western culture and from Eastern to Western cultures are associated with a substantial loss of bearings that normally allow a person to make sense. This transnational migration brings about a shift in identity that is also experienced in learning science, where students are introduced to the new languages of the subject matter by means of everyday language that is of the dominant culture of the school or university. Locating this chapter in their own experience of moving between nations and cultures, the authors articulate issues that the shift to speaking a language other than their mothers' tongues brings forth to the experience of self and how it mediates learning in and of science.

Another form of cultural relation is that between hip hop and standard culture, itself already a hybrid arising from the continual bricolage that occurs at the interstices between the middle-class values underlying schooling (Eckert, 1989) and the different cultural roots that characterize students' lives outside schools. Although we can think of hip hop as a culture within culture, we ought not to essentialize the phenomenon but rather understand it as multicultural at its heart. In chapter 5, "Reality Pedagogy: Hip Hop Culture and the Urban Science Classroom," Christopher Emdin shows how hip hop serves students of color in urban areas as an escape from the struggles of their everyday lives. Lyrics to rap songs tell tales of both the physical realities of life in the inner city and the emotional frustration that comes with being ostracized from and silenced in mainstream culture. There is a mutually constitutive relationship between rap and the inner city that those that are involved in hip hop deeply understand. The streets speak to the music and the music reports what it hears from the streets. Those individuals not actively involved in hip hop often believe that the hip hop generation is the underbelly of American culture. These individuals fail to realize that hip hop is a product of a lack of voice in schools and the political arena. In his research, Chris finds urban students engaged in hip hop culture to possess many attributes that support success in science that are not fully explored in urban classrooms. He constructs for us a path to uncover students' experiences that are integral to the teaching and learning of science. He achieves this goal by engaging in dialogues with students who are both participants in hip hop culture and students of science in secondary schools. Through these dialogues, Chris presents a reality-based urban science pedagogy nested in students' experiences and hip hop culture.

Diaspora is an old phenomenon and, more recently, has become an important theoretical concept in cultural studies generally and in science education specifically (Roth, 2008a). It allows us to understand the experience of migrants between different parts of the world. In the US, there is a new phenomenon whereby rural communities are rapidly becoming unofficial sister cities to rural communities in Mexico. Scholars denote this phenomenon by the term "New Latino Diaspora" (Wortham, Murillo, & Hamann, 2002). Small-scale Mexican farmers and their families, displaced due to North American Free Trade Agreement-induced privatization of once-communal farmland and a flood of cheap American corn, are moving north in record numbers to secure employment in the de-skilled and de-unionized meatpacking industry of the American Midwest. This diasporic movement from the south to the north involving nearly entire communities constitutes an historic transformation in the labor markets and living and learning modalities of both sending (Mexican) and receiving (Iowan) rural communities. The struggle of small independent farmers against the industrialization and centralization of agriculture is a unifying theme on both sides of the border. Yet the heated rhetoric over illegal immigration and the "browning" of America's heartland drowns out this common concern.

Drawing from a multi-site ethnography of two Mexican and Iowan sister cities, Katherine Richardson Bruno and Hannah Lewis contrast in their chapter "Sister City, Sister Science: Science Education for Sustainable Living and Learning in the New Borderlands" the coordination of "work knowledges." This coordination cobbles together and transforms the knowledge Mexican families bring to their participation in the U.S. food production systems with the canonical content knowledge of school science. Challenging the long-standing subjugation of these everyday work knowledges in mainstream science, Katherine and Hannah argue instead that such knowledges are critical to reconceptualizing science education in the age of globalization and global migration. The current privileging of U.S. economic advancement through technological innovation, offered as a rationale for efforts to improve science education for non-dominant students in the reforms of *Science For All*, does not speak, for example, to the desire of immigrant students and their families to return to the farming lives they left behind in Mexico. Nor does it speak to them should they remain in the US and move away from the exploitative labor of packing plant work to be able to work where their agrarian backgrounds support what they do. Taking the goal and framework of *sustainable rural livelihoods* as its point of departure, the chapter concludes by re-framing science education from a globalized and diasporic standpoint and outlines the need for and nature of a proposed border science curriculum for transnational living and learning. Such a curriculum, it is suggested, could be a source of social dreaming and healing for all students and families in the sister cities of the new borderlands.

In chapter 7, the authors of the preceding chapters discuss implications that spring forth from their work organized around four questions that the editor asked them to respond to. In this *metalogue*—according to Gregory Bateson (1972) a conversation that takes learning to a new level by learning about learning—the authors discuss issues such as (a) the divide between academics and everyday folk, (b) teaching science in ways that respect everyday forms of knowing, (c) the possibilities that come with everyday knowing and place-based science education, and (d) the relation of teaching "authentic science" and everyday knowing.

Part II: Othering the Self, Selfing the Other

Despite the traditional rhetoric of making science relevant to students, their lives and life experiences are generally excluded in the quest of inculcating (allowing self-construction of) the "right" scientific knowledge as specified in national standards. Even the staunchest constructivists, claiming that knowledge is personally constructed on the ground of existing knowledge and understanding, nevertheless want their students to arrive at the "right," "canonical" form of knowledge. In the form of conceptual change theory, constructivist educators actually aim at rupturing students' existing understandings, attempting to make them "restructure" their mind from misconception or alternative conception to the correct, scientific conception. The ways of the students' homes and everyday lives thereby come to be devalued and students are asked to abandon forms of discourse that continue to have currency in their lives outside the science classroom. Not surprisingly, then, we find that only a small percentage of students like science and pursue science-related careers after graduating from high school.

All three chapters in this section and the metalogue that follows allow us to see the active interplay that exists between Self and Other (the generalized other), each of which presupposes the respective other. In constructivist terms, the Other is a figment of the Self, the result of a construction tested for its viability in exchange with whatever the term denotes. The Other thereby is made in the image of the Self, which is the source of the construction and which thinks the other in its own image. This framing of the self–other relation in educational theorizing at the end of the 20th century is surprising given that, already at the beginning of the century, the philosopher Edmund Husserl showed that a self could never construct anything like an other. Thus, Husserl realized that "I" cannot identify the behavior of someone else as angry or wrathful without first adopting the viewpoint of another on my own affects (Franck, 1981). It is under this sole condition that the "I" can recognize another's bodily manifestation as indicating anger or wrath. It is therefore not the Self that serves us as a model for the Other, but rather the Other that serves as a model for the Self: anything we can express about language—which we appropriated from the other, for the purposes of communicating with the other, in a process of which language returns to the other—inherently *is* other than the Self.

At the same time, an agential conception of the Self introduces the possibility to produce novel expressions, which, though already circumscribed within the currently possible language, nevertheless is realized in a singular way and therefore produces resources that become available to the other, who thereby comes to shape himself or herself in the image of the Self. Both Self and Other, therefore, are impure, metis, always and already utter singularity and absolute general at the same time.

The chapters in this section show that science, scientific knowledge, and scientific literacy—if these terms are to be relevant to everyday people coping in their everyday lives—can and ought to be rethought from the perspective of the individual. The authors in this part take their own personal experiences—Angela Calabrese Barton as expecting mother, myself as a chronically ill individual, and Karen Tonso as a woman in engineering—as a starting point for a critical interrogation of science and scientific knowledge for developing ways of thinking about science education.

An integral aspect of many families generally and of women specifically are the times of pregnancy and the early years in a child's lives. Families are struggling with the situation and they strive to know more about how to deal and live with the soon-to-come or recently arrived. The needs to become knowledgeable about health, illness, and so on are salient, and the question we ought to pose regards how we can rethink science to be relevant in such situations. This is the topic of Angela Calabrese Barton's chapter "Mothering and Science Literacy: Challenging Truth-Making and Authority through Counterstory." When she became pregnant with her first child, Angela was told by many people: "You need to read *What to expect when you are expecting.*" So, as a dutiful new mother she went out and purchased the book, and about five others describing the ins and outs of pregnancy and babies. After all, she felt like a science educator and knowledge of how the body works is interesting to her, especially when it is her own body! But Angela quickly became frustrated because the tone of most of these books not only felt paternalistic, but also essentializing, as if all pregnant bodies worked the exact same way. Angela knew she should feel some morning sickness and should sleep with crackers next to her bed to eat when she woke up to alleviate my symptoms. She should not exercise rigorously and should stay away from non-healthy foods like ice cream. She did understand the "science" behind these recommendations, but the essential claims did not fit her world. The thought of crackers made her stomach churn, and running an easy 4 miles made her feel well—as I know from having met her after a run while she was in her last month of pregnancy. So, she expanded her search for information, including joining an on-line community of other about-to-be mothers. A subset of them formed their own private forum because, as they got to know each other better, they wanted a safer space to share their stories and experiences about their children and themselves. Since that time many of these mothers have met "in real life" though their friendships survive on-line.

The problems in the early life of her second baby emerged in this context of her early experiences. Whereas this community has served as multifaceted site in terms of the reasons why the women post there, as a mother who is interested in science learning and science literacy, Angela was struck at how this space has become a knowledge-generating community that lacks both the essentialist and paternalistic overtones of those books she purchased five years prior to that time. The knowledge of this community, which is distributed, personal, contextual and often contested, has not replaced the world of doctors or formal medicine, but has become one of the filters Angela uses to understand herself and her children's health. For example, beginning from birth, the head of her second child always tilted to the right. Her pediatrician did not think much of it because at the doctor's office her daughter never quite "performed" in a way that would demonstrate this tilt. But it nagged Angela and her partner. She took a few pictures and showed them to her on-line friends. Two immediate responses in particular suggested that this looked like *torticollis* (a Latin word that literally means "twisted neck"), and the writers suggested that Angela ought to pursue it before it caused head bone misalignment. As it turns out, each of these women has had a child with *torticollis*, but neither child was diagnosed on time and both subsequently required the corrective helmets to realign the head bones. Indeed, 5 minutes in a pediatric neurologist's office confirmed this diagnosis for Angela's daughter, and she is grateful to her on-line friends for their early intervention! Fortunately for her daughter, corrective physical therapy allowed the body to heal itself. In her chapter, Angela explores how this on-line community of mothers without medical degrees (and many without college degrees) has provided her with a collective wisdom that is personal, contextual and contested.

In chapter 9, I use my own experience of living with chronic illness to reflect on science and science education. Chronic illness has become salient in my life, as I have become increasingly aware in recent years that there are many people in my wife's or my own workplaces who live with chronic (sometimes terminal) illness. After two bicycle accidents in 2001, I found myself physically and cognitively impaired, without initially linking the accidents and my state. There followed years of testing for different kinds of possible illnesses, including amyotrophic lateral sclerosis (ALS or Lou Gehrig's disease). In all of this, I found myself at the mercy of a system that did not and perhaps could not help me, leaving it up to myself to live through a frequently debilitating condition that ultimately received a name: chronic fatigue syndrome/fibromyalgia. In chapter 9 entitled "Living with Chronic Illness: An Institutional Ethnigraphy of (Medical) Science and Scientific Literacy in Everyday Life," I use this experience and my search for a scientific understanding of what was happening to me and of the solutions I envisioned and enacted, which ultimately were associated with a radical improvement of the condition. The experience and my understanding thereby constitute the ground for understanding and redefining science and scientific literacy in the everyday life of a person who not only is led to coping with the situation but also to evolve ways of mobilizing science and scientific research reports to provoke a return to wellness.

For 19 years, Karen L. Tonso was an engineer after having made it through a male-dominated training and into an equally male-dominated profession. After deciding to return to graduate school to become an educator, she found an opportunity to revisit her own training in the light of an ethnographic study in a public engineering school. In "A Stranger in a 'Real Land': Engineering Expertise on an Engineering Campus" (chapter 10), Karen revisits the question of what it might mean to become and be an expert. In the mid-1990s at Public Engineering School (PES) in the U.S. mid-continent, two forms of engineering expertise existed. One form aligned with an academic-science form of life associated with conventional, ABET[2]-accredited curricula in engineering education, the other embraced a comprehensive set of understandings and practices, in my experience and according to studies of engineering, better suited for work as "actual" engineers, student engineers' term for practicing engineers. The second form of expertise emerged from reform efforts, especially appending a design curriculum to conventional curriculum, where students worked in teams to complete projects for industry and government clients. With 15 years of industry engineering experience, it took her little time to recognize "real" engineering expertise when she saw it during student teamwork, but it has been quite another matter to convince engineering educators that *academic-science expertise*, privileged at PES (and other campuses) by campus traditions, routines for success and excellence, teaching practices (especially learning activities and grading), and other cultural norms will in fact be a "stranger" when students enter industry careers, or figuratively the "real" land. This chapter juxtaposes her perspectives as educational researcher and as former engineer to reflect on and trouble engineering expertise, and suggest why the one preferred at PES is arguably "strange" to a "real" engineer. Doris Lessing's observation about the elderly—"Your body changes, but you don't change at all. And, that, of course, causes great confusion"[3]—captures the nature of reform at PES.

In chapter 11, the three contributors to this section engage in a metalogue that covers emotions, knowledgeability, the relation of multiple knowledges as source for decision making, and the contradictions that science teachers and science educators face when they are to take into account the personal knowledges people evolve in the course of their lives, on the one hand, and the frequently incompatible scientific knowledges that are to be taught in the schools, on the other hand.

Throughout the book, it becomes evidence that taking into account peoples' lives means dealing with and *appreciating* differences, especially those that arise from the contrast and contradictions between science and everyday ways of knowing. I therefore end this book with an epilogue in which I reflect on difference as such, that is, in and for itself, and how to appreciate difference when we think about and plan curriculum.

Notes

1 4-H is a youth organization in the USA, administered by the Cooperative State Research, Education, and Extension Service, which allows young people to develop citizenship, keadership, and life skills through programs based on experiential learning.
2 ABET stands for Accreditation Board for Engineering and Technology and is the recognized accreditor for colleague and university programs in applied science, computing, engineering, and technology.
3 http://www.worldofquotes.com/author/Doris-Lessing/1/index.html

References

Bakhtin, M. M./Vološinov, V. N. (1973). *Marxism and the philosophy of language.* Cambridge, MA: Harvard University Press.

Bateson, G. (1972). *Steps to an ecology of mind.* New York: Ballantine.

Eckert, P. (1989). *Jocks and burnouts: Social categories and identity in the high school.* New York: Teachers College Press.

Epstein, S. (1995). The construction of lay expertise: AIDS activism and the forging of credibility in the reform of clinical trials. *Science, Technology, & Human Values, 20,* 408–437.

Franck, D. (1981). *Chair et corps: Sur la phénoménologie de Husserl* (Flesh and body: On Husserl's phenomenology). Paris: Les éditions de Minuit.

Nancy, J.-L. (1993). L'éloge de la mêlée [Eulogy to the mêlée]. *Transeuropéennes, 1,* 8–18.

Pauly, D., Pitcher, T. J., & Preikshot, D. (1998). *Back to the future: Reconstructing the Strait of Georgia ecosystem.* Vancouver, Canada: University of British Columbia Press.

Rabeharisoa, V., & Callon, M. (1999). *Le pouvoir des malades: L'Association française contre les myopathies et la recherche* (The power of the ill: The French Association against Dystrophies and research). Paris: Les Presses de l'école des Mines.

Roth, W.-M. (2008a). Bricolage, métissage, hybridity, heterogeneity, diaspora: Concepts for thinking science education in the 21st century. *Cultural Studies of Science Education,* 3 891–916.

Roth, W.-M. (2008b). Constructing community health and safety. *Municipal Engineer, 161,* 83–92.

Roth, W.-M., & Barton, A. C. (2004). *Rethinking scientific literacy.* New York: Routledge.

Smith, D. E. (2005). *Institutional ethnography: A sociology for people.* Lanham, MD: AltaMira Press.

Wortham, S., Murillo, E. G., & Hamann, E. T. (Eds.). (2002). *Education in the new Latino diaspora: Policy and the politics of identity.* Westport, CT: Ablex.

Part I

Culturing Knowledges

Introduction to Part I

> Every culture is in itself "multicultural," not simply because there always has been an antecedent acculturation but more profoundly because the gesture of culture itself is a gesture of the mêlée.
>
> (Nancy, 1993, p. 13)

In each chapter of this first part of the book, multiple cultures come to confront each other. Eileen Parsons tells the story of two African American women who, in different ways, come to face science, which is by and large the cultural achievement of white male middle-class scholars—Jacques Derrida coined the term phal-logo-centrism to denote this male-dominated dimension and cultural heritage of scientific thought and process rooted in the Greek idea of logos. Carol Brandt grounds her chapter in her experiences of agricultural practices within a Midwestern German-American community that come to be confronted with the farming practices at Zuni Pueblo. SungWon Hwang and I use the experience of immigrating to Canada and studying in a language that was not our own to think about a double confrontation of culture/language, on the one hand, and of everyday understanding and science, on the other hand. Chris Emdin focuses on hip hop, a culture within a culture that comes to be confronted with another facet of the same but nevertheless different culture, the dominant Anglo-Saxon, white culture characteristic of schools and science. Finally, in Katherine Richardson Bruna and Hannah Lewis's chapter we find Mexican immigrants, who bring a lot of manual farming experiences and practices with them into their new (temporary) homeland, confronted with industrialization and deskilling.

In the educational literature, encounters between different cultural forms have come to be theorized in terms of the concept of "third space," which is used to denote (both official and unofficial) literal and metaphorical distinct and exclusive spaces where teachers and students interact (e.g., Gutiérrez, Rymes, & Larson, 1995). But used in this way, the concept reifies a form of thinking culture that was not what the inventor of the concept Homi Bhabha (1994) articulated in *The Location of Culture.* In his use, the concept is a dialectical one, always and already in existence with each utterance that brings about a shift between the speaking subject and the subject of speech. This shift is a form of diastasis, dislocation, shear, that is, a shift within an entity that comes to be dislodged within and with respect to itself. But a much better articulated approach to within and cross-cultural difference comes from continental, largely French philosophy that thinks difference as such and takes it (i.e., difference as such) as the starting point for theorizing culture (e.g., Deleuze, 1968/1994).

Cultures are non-self-identical entities, i.e., multiplicities, that further hybridize themselves in each action that produces and reproduces them. Even if the same word is uttered twice within the same sentence, it means something different the second time because it is now heard or read with reference and in relation to its first appearance (Bakhtin, 1986). When, by being heard or read, the word returns to the other, it means something different again, thereby leading to the hybridization of sense and personal meaning each time it or something related to it is uttered/written. When aspects and members of two cultures come face to face with each other, such as when Mexicans come to make a living in the US or students from an Asian country migrate to North America in the pursuit of a university degree, new forms of an already multiplicitous, plurivocal, and polyphonic culture emerge. Rather than occurring in some third space, where first and second cultures are taken as purities, cultures are the heterogeneous locations where the mixing occurs. These mixing processes, which begin with mixtures that are mixed to ever-increasing degrees, constitute interesting models for rethinking culture. Here, I propose the term *mêlée* to denote a continual process in which hybrid processes further hybridize each other. I draw on the following brief story and analysis to exemplify this idea.

In a laboratory office of a midsize Canadian university, an Indian scientist (pseudonym Macarthur) and a Chinese graduate student (pseudonym Yi) have come together for an interview about the biography of the former. They speak (fluently) English, though for both it is their second language. In the course of their talk, they produce an (audiotaped) narrative, in which the plot of Macarthur's life is a mixture of the forms that the autobiographies of people in Western societies can take: against the goals his parents have had for him, he chose his own field, went abroad to Canada to get his PhD, and eventually came to build and lead a world-renowned laboratory. In this life plot, the person Macarthur came to be a particular character—the one that marked the end of the plot with his current life—in part on his own, in part as a function of the plot. As a result—because the interview requires the collaboration of interviewer and interviewee—the topics covered, the characteristics talked about, the features of Macarthur the person, his character and life trajectory, came to be a mix, or, as I propose, a momentary image of a continual mêlée of culture. In the story that the interview situation produced, his life trajectory could not just take any singular form but always and already had to take a form that was intelligible to others and therefore to himself. The life story of Macarthur, to be intelligible, had to be told in the language that had come to him and Yi from the Other to which it returns, by means of this and other published narratives. The auto/biographical account Macarthur and Yi produce, therefore, is simultaneously singular, being told there in a non-repeatable way about a non-repeatable life, and an instance of a genre, a way of telling a life. Macarthur's life and his identity come about through the mêlée of all possible utterances and voices that are refracted in and through the interview text.

I have taken the concept of mêlée from an essay by the French philosopher Jean-Luc Nancy (1993), who uses it to theorize a number of phenomena such as culture, language, knowing, subject of knowing, and so forth. The concept of the mêlée highlights the process character of these terms and the things they denote as processes of mixing and mixes rather than purities; and I intend this process of mixing to be associated with all the connotations of the word including scramble, scrimmage, batter, and scrummage:

> in a mêlée, there is opposition and encounter, there is what gathers itself and that which separates, that which makes contact and that which makes contract, that which concentrates and that which is spread out, that which identifies and that which alters.
>
> (Nancy, 1993, p. 12, my translation)

Thus, in the interview, which Macarthur and Yi produce with the same kind of skill that characterizes jazz musicians improvising during a jam session, neither the interviewer Yi nor interviewee Macarthur can be understood as self-identical unities (stable identities). Rather, their identities, in a contradiction of terms, always and already have to be understood as mêlées, continual processes of change. The language they use, English, also is a mêlée, a continual mixing of languages, even of the language with itself—and it therefore is in constant change, continually mixed and remixed in use (Bakhtin, 1986). An upshot of this view is that we have to rethink scientific language and literacy and the forms of identity that students evolve relative to science as they participate in school science.

Jean-Luc Nancy had written his praise of the mêlée to articulate issues of culture, language, and identity in war-torn Sarajevo of the 1990s, though he generalized his comments to culture, language, and identity as such. I propose the mêlée as a concept for thinking such educational issues as scientific culture, language, literacy, and identity. I do so not to reproduce the concept in carbon-copy form—which I could not in any case because of the continually changing sense of words (Bakhtin/Vološinov, 1973)[1]—but to develop an educationally relevant concept that has fractal similarity with the way in which the philosopher uses it for his quite different purposes (understanding the situation Sarajevo). This concept is built on the grounds of an ontology that theorizes difference to exist in and of itself. Any phenomenon (language, Self, culture) is taken to be non-identical with itself and therefore has difference as the core constituent of identity ($I \neq I$). Difference at the core of the Self (same) is able to harbor the possibility of the impossible—for example, the attainment and learning of that which is absolutely foreign/strange and therefore lies beyond the horizon of the intelligible. Thus, scientific literacy, scientific knowing, and science identities already come to be possibilities at the very heart of what they are not: pre- and un-scientific literacy, non-scientific knowing, and non-science identities.

Thinking of language as a mêlée in the way I articulate it here, as a continual process of mixing, has important implications for theorizing and enacting science in the classroom. The language students are to learn—the one that characterizes them as scientifically literate in the eyes of science educators and scientists—has to be acquired in the form of translations from the language that they currently speak, a language many science educators have come to characterize as embodying misconceptions, alternative frameworks, preconceptions, and other deficit phenomena. Whatever the teacher says in a scientifically correct and consistent way can be made intelligible only through a *translation* into the vernacular, which is different in its structures, timbres, and voices but, in the course of instruction, assumed to be the same. It is the same English or French, but also a different English or French. It is the same and different simultaneously, for when a teacher uses it, it is scientific, and when students use it, it is characteristic of misconceptions and alternative ways of speaking. It is English (French), but it is a different English (French): the English (French) language is not identical with itself. There are as many forms of English as there are speakers speaking it, and every single speech act changes it. Yet these different ways of speaking are the very (fertile) ground *on* and *from* which scientific ways of speaking emerge and evolve by means of translations that mean the same although any two expressions that are made to correspond are different. Any single language therefore already is a mêlée of languages (see Nancy's quote introducing this section), and one of these other languages is the language itself (Derrida, 1998). Translation therefore "translates itself in an internal . . . translation by playing with the non-identity with itself of all language" (p. 65).

The concept of mêlée, because it is built on and consists of difference, allows for individual styles, tones, voices, genres, and so on to exist and emerge within a culture; but it also allows for the styles, tones, intonations, voices, genres, and so on that are required for the original to be heard, understood, and interpreted. Thus the sciences constitute particular styles, tones, intonations, voices, and genres; but these arose on a phylogenetic level, and, on an ontogenetic level, continuously arise, from everyday ways of talking that pre-existed and pre-exist them. As Bakhtin/Vološinov (1973) suggested, changes in high literature over the course of historical time *cannot* be understood by deriving change from high literature itself. Rather, changes in high literature are reflections of continual changes produced in each utterance, where the authors understand each novel, poem, or other literary work as but another utterance spoken in the culture at large. This is an important step in theorizing learning, for in science classrooms common life comes face to face with the practices of particular communities, cultures within a culture, with practices (if we follow the distinction between scientific and pre-scientific and non-scientific) that are different from the remainder of the culture.

When reading about the two African American women in Eileen Parson's chapter, about the Zuni and Midwestern cultures in the chapter by Carol Brandt, about the different cultures involved in transnational migrations, hip hop and mainstream (Emdin), or the encounter between Mexican and heartland American agricultural discourses in Katherine Richardson Bruna and Hannah Lewis's chapter, we ought not think in terms of the traditional oppositions and comparisons. We ought not think in terms of how different these cultures are from each other but in terms of how different they are from themselves and within themselves. It is only when we understand difference as such, when we accept thinking in terms of difference that we come to understand the problematic nature of difference as constructed by researchers, official (political) discourses, and by the reigning ideologies that constitute the ruling relations in rural Iowa, part of the U.S. heartland.

Once we understand the polyphonies of the texts that come from our own mouths and keyboards, we come to appreciate cultural difference and sameness in new ways. We will no longer celebrate the third spaces in which our students are said to cobble together new forms of culture, but come to understand that our everyday cultural practices are themselves forms of bricolage that continuously produce and reproduce difference in a mêlée of forms and content where difference comes to brush up against itself. When Carol Brandt learned the Zuni way of gardening, her own patterned actions—although they had always been changing and becoming better, but more suited and suitable to the American Midwest—now continued to change more drastically in the encounter and became more suited and suitable to gardening in the arid landscape of the Zuni lands.

Above all, we need to keep in mind: purity *is* not. But neither *is* there impurity. There only is the continual process of mixing, reflexively mixing the processes of mixing. The accounts provided in the following chapters allow us to see this mixing *while it occurs*, in different geographical parts of the North American continent: Eileen Parsons' Southeast, Carol Brandt's Southwest, Hwang *et al.*'s Pacific Northwest, Chris Emdin's inner-city New York (Northeast), and Katherine Richardson Bruna and Hannah Lewis's heartland.

Note

1 There is an ongoing debate about who wrote the book. In Anglo-Saxon translations, the work is generally attributed to V. N. Vološinov (there are exceptions), whereas in Portuguese, French, and German translations, the authorship is attributed to M. M. Bakhtin with Vološinov as a contributor. I prefer the French to the English version because it does better justice to the Saussurian heritage of the works (e.g., the distinction of sense and reference rather than meaning).

References

Bakhtin, M. M. (1986). *Speech genres and other late essays.* Austin: University of Texas Press.

Bakhtin, M. M./Vološinov, V. N. (1973). *Marxism and the philosophy of language.* Cambridge, MA: Harvard University Press.

Bhabha, H. K. (1994). *The location of culture.* London: Routledge.

Deleuze, G. (1994). *Difference and repetition* (P. Patton, Trans.). New York: Columbia University Press. (First published in 1968.)

Derrida, J. (1998). *Monolingualism of the Other; or, The prosthesis of origin.* Stanford, CA: Stanford University Press.

Gutiérrez, K. D., Rymes, B., & Larson, J. (1995). Script, counterscript, and underlife in the classroom—James Brown versus Brown v. Board of Education. *Harvard Educational Review, 65,* 445–471.

Nancy, J.-L. (1993). L'éloge de la mêlée [Eulogy to the mêlée]. *Transeuropéennes, 1,* 8–18.

2 Revisiting and Reconsidering Authenticity in Science Education

Theory and the Lived Experiences of Two African American Females

Eileen Carlton Parsons

The term *authenticity* is used in various areas and in different ways in science education. Among the many domains, the term can be found in investigations and discussions on assessment, curricula, educational contexts, and instructional approaches like inquiry and project-based learning. In each of these areas, the meaning of *authenticity* is seldom explicit. Oftentimes, readers must infer its meaning from how the term is employed and how it is situated within the surrounding texts. In this chapter, I revisit and reconsider the notion of *authenticity* via a parallel presentation. One exposition illustrates the previously stated in the life spaces of two African American women whose experiences are situated in two distinctive eras in the history of the US. The other presentation examines *authenticity* from a theoretical perspective. I conclude by positing a multidimensional view of authenticity in science education for African Americans as a collective.

LIVED EXPERIENCES OF TWO AFRICAN AMERICAN WOMEN: 1927–2007

I feature two African American women born and reared in the Southeast region of the US. The ages of these two women are separated by half a century, a span marked by significant historical events pertaining to racial equality. The life spaces of these two women are vastly different in some ways, but strikingly similar in others. Even though I highlight the lives of these two women, I, as the narrator, am subliminally present in the (re)presentations of their stories and in the positioning of their stories with respect to authenticity. My life space is also reflected to varying degrees in the very stories of these two women.

REVISITING AUTHENTICITY IN SCIENCE EDUCATION

I briefly review authenticity in the science education literature and then revisit it from a perspective not typically employed. The reference point from which I revisit authenticity is the triple quandary-superimposed (TQS) cultural-historical activity theory (CHAT)-Ecological framework (Parsons, 2008).

The Meaning of Authenticity in Science Education Literature

A recent review of science education literature that utilized the term "authentic" or "authenticity" subsumed

The elder African American woman was and still is an influential figure in the church and community in which I grew up. Singing Negro spirituals at church and worship events; preparing young adults and teenagers to serve respectfully in the functions of the church; and directing children to act appropriately within the holy sanctuary are among my earliest memories of Ms. Herlton (a pseudonym). Her story captures many of the experiences of my grandparents and parents; subsequently, her life experiences contextualize and sketch the formative years of my life as one growing up a part of the African American, poor working class in the rural southeastern US.

I met the younger African American female in a life space that is far removed from the life space in which Ms. Herlton exerted her life-impacting influence. I encountered LaShaundra (a pseudonym) in my role as a science education researcher. The alignment among our research interests in social justice, African Americans, and science education resulted in mutually supportive professional and personal relationships. With regards to her road to educational achievement, many facets of LaShaundra's story resemble my own. Though distinct and divergent in many respects, the lives of Ms. Herlton and LaShaundra, and my life spaces as the author, converge in the notion of authenticity I present in this piece.

Ms. Herlton is an 80-year-old female who currently earns an income by baking and selling goods (e.g., apple pies, breads, cakes). She has resided in the same impoverished, rural region her entire life. She lived her childhood, adolescent, and young adult years in a segregated society where African Americans were legally relegated and subjected to second-class citizenship. Because she needed to work so her younger siblings could go to

the operative notions of the said terms under three categories (Buxton, 2006): canonical, youth-centered, and contextual. The canonical classification includes literature that describes authenticity with respect to what occurs in Western science and what is promoted in widely acclaimed documents like the *National Science Education Standards.* That is, the canonical view equates authenticity to scientific knowledge and scientists—their dispositions, outlooks, and practices—working within scientific communities; the more closely practices resemble those of scientists the more authentic. National concerns like the preparation of more US-born scientists and a scientifically literate citizenry capable of contributing to a technologically sophisticated and global society are among the overarching outcomes in the pursuit of authenticity from a canonical position. The perspectives, interests, desires, needs, and practices of students and communities that involve them are the origins and essentials of authenticity from a youth-centered perspective, the second category. More specific than the primary aims of the canonical view, to help individuals understand the nature of science and the scientific enterprise and solve real-life problems are among the goals of the youth-centered perspective. The third classification, contextual, is a hybrid of the first two categories. Tenets and goals of the canonical and youth-centered categories constitute the contextual view of authenticity. Summarily, in the contextual classification, the desires, perspectives, interests, and needs of students are the means through which the ends—what transpires in scientific enterprises and the resolution of context-specific problems—are obtained.

school, her formal education ended after the ninth grade.

Unlike the elder Ms. Herlton, LaShaundra experienced many years of schooling. LaShaundra, a 30-year-old female, grew up in a predominantly European American, middle-class, rural neighborhood and attended schools populated largely by European American students. She left the rural area to attend Predominantly White University (PWU), located in an urban area where she now resides. She holds a BSc degree in meteorology, and MSc and PhD degrees in science education. A society where African Americans and European Americans are not legally segregated is the society she knows. Through the life stories of Ms. Herlton and LaShaundra, notions of authenticity relevant to science education emerge.

Back Then and Now: The Life Spaces of Ms. Herlton

It was a hot, humid afternoon in the summer of 2007. A sweet-sour aroma of simmering fresh apples filled the air and the hum of a television slowly vanished from my consciousness as I listened intently to Ms. Herlton. The toils and tolls of time upon her sturdy physique were evident by the sling that hoisted her right arm and the slide and shuffle of her walk. But her voice was strong and unwavering as she talked about her life and shared words of wisdom.

"I guess I started cooking at six years old. Most of the time, until I got up to a pretty good-size girl, I stayed at home and took care of the children and did the cooking. And when we got up so we could go out, the girls—my sisters—there was five of us girls and one boy, and we would go out into the woods and skin black haw bark, dewberry root, schomac, and all. We'd gather herbs—star root, mountain tea, and all that. My

Buxton's (2006) review answered the question "What does authenticity mean?" a question other scholars also pose. These authors also raised an additional query of importance when considering authenticity: What does authenticity mean according to whom? In this chapter, I pose a third question: authenticity for whom and for what purposes? I employ the TQS CHAT-Ecological Model in contemplating the nature and significance of the previously stated questions.

Components of the TQS CHAT-Ecological Model

TQS CHAT-Ecological Model resulted from the synthesis of three distinct conceptual frameworks posited by psychologists. In capturing the tensions experienced by African Americans residing in the US, A. W. Boykin (1986) posited the triple quandary. Michael Cole (1996) developed CHAT to highlight and explain the significance of culture in human activity; and Uri Bronfenbrenner (1979) constructed the ecological systems theory to emphasize and explicate the interconnectedness of various environments in human development. The synthesized construction, the TQS CHAT-Ecological Model, offers a heuristic for acquiring a comprehensive understanding of an individual's experiences as separate from and a part of a collective in milieus in which stratification occurs.

The Triple Quandary

The African American experience in the US is represented in the triple quandary as three interacting and contentious daddy would go look at a place. Like if they were cutting timber, he would go look at a place. After he looked at a place then he maybe go look at a field that didn't have no corn or nothing growing in it. Well, Lobilly [Lobelia] growed in that so we [Ms. Herlton and her sisters] would pick and pull Lobilly out, cut the roots off, dry it and carry it in. If they [her father and others] cut poplar trees we would pull the poplar leaves and dry them. But you had to get them dry 'cause if you packed 'em in and they wasn't dry enough they turned black. You had to make sure they was dry. We would use 'em and sell 'em.

"We used dewberry root to make a tea that treated diarrhoea; Lobilly to treat bronchitis and arthritis; and the poplar leaves to make a hot poultice that broke a fever. A tea was made out of the yellow root to lower high blood pressure and for yellow jaundice. The mountain tea, blackberry root, or peach tree leaves were used as teas for stomach aches.

"They had a thing called danagebud—I don't see no trees now. But it growed up as big as your thumb. You tie it and get all the leaves off and they would buy it green. You pick it and haul it in. Then there were the briar roots. You know, you had to cut the briars off of that. But, like they had a sawmill here. This year before they got through [working in the sawmill] the briars would begin to grow up so the next year we would go and pull the briar roots, bring them home, beat the briar roots, dry 'em and then sell 'em.

"And poke root, you've seen poke root. 'Cause they grow up most anywhere. They have a red thing on it and if you pulled it off they get you red. Right now, they begin to turn red and you can take that and make wine out of it. And then you see, you have the pokeroot, the blackberry, and the spheres—oppression, African-rooted culture, and mainstream U.S. culture. The triple quandary is based upon two premises.

First, the triple quandary assumes commonalities among a group of individuals who share a socio-political past and present. The acknowledgment of these commonalities does not lessen the value and uniqueness of individuals who are members of groups and does not imply that these commonalities are evident in group members' lives in the same way and to the same extent. The considerations of commonalities implicitly implicate the contexts in which a group functions and the impact of contexts upon the agencies of individuals who are group members.

The second assumption of the triple quandary builds upon the group commonality premise. Within a society in which groups are devised according to and stratified by socially constructed identifiers such as social class, race, and the like, value is often assigned to the various strata. Consequently, material (e.g., wealth) and symbolic (e.g., power) resources are differentially accrued in discernible patterns across strata. Subsequently, groups of dissimilar statuses across strata perceive and live substantively different realities. Like the old adage of the blind man who interprets experience based upon the part of the elephant he is able to perceive through touch, the realities of various groups living in a stratified society are influenced by the places they occupy in the stratification.

The primary assertion that emanates from the two premises of the triple quandary pertains to the specific positioning of African Americans in the elderberry—they don't grow no more. Because they've sprayed, they have sprayed and killed them out, but you pull them [the poke root and the berries] off and you use that. Say you sold them. Say you had 50 pounds, well all you had was 50 cents because they only paid you a penny a pound. Like, the pine—that [part] that is down at the bottom. It roses and comes out white and pretty. But that [part] that goes up there where the limbs are up high, you can sell that. Just skin that and dry it. Make sure it is dry and you can bale that and they'll buy it like that. I don't even know if they're buying that now but I do know that they're buying poplar bark and the logs.

"Then after I got to be 12 years old I went out into service to work for different people. I went to a boarding house, worked for a doctor, different people I would work for. I only made $2.50 a week. When I got that done, they'd give me a raise, maybe I get $3. And if it was raining, like all week it was raining, then Daddy would meet me down there and get the $3 for the family. And like if we cooked at the boarding house, if we cooked a ham, well, I wouldn't throw away the scraps. I would save that for Mama to have for the family. All we had, well, you see we had blackberries, our own cow, our own milk, beans, and peas. We picked beans and peas and peaches and apples, and we canned that. We dried our own apples and we did our own canning. Picking blackberries and canning them, peaches and all that. We did all that for ourselves. We didn't go to the store to buy that. We only bought salt. We would go to the mill and take our corn, and go there and get our meal made. We made many things for ourselves.

"Now you can't, with what I do [sell apple pies], you can't make pies out of the apples you dry. It's a health code, if you make apple pies out of them [dried

US. Both the historical and contemporary representations of African Americans as a collective in the social, political, economic, and educational arenas are evidence of the group's low station. For example, in relation to the group's percentage in the U.S. population, African Americans are over-represented in low-status positions (e.g., prisoners) and under-represented in high-status ones (e.g., judges). Many conjectures (e.g., genetic endowment) are introduced and discussed in the literature as contributors to the inferior positioning of African Americans in U.S. society. The triple quandary highlights two of these propositions—racial oppression under the label "oppression" and cultural hegemony, a relation captured in the realms of "African-rooted culture" and "mainstream U.S. culture." The triple quandary situates racial oppression and cultural hegemony as two major determinants of the low status of African Americans.

I define racial oppression as the systematic, prolonged, and race-based denial of a group's access to material and symbolic resources that have the potential to advance the group within the power structures of society. Ms. Herlton and LaShaundra share examples of racial oppression in their life stories. Although Ms. Herlton grew up in an era when society approved and legitimized racial oppression and LaShaundra knew only a desegregated racial society, their examples highlight the way in which access for African Americans is systematically denied and restricted. For instance, Ms. Herlton implicated the segregated society in recalling her early work experiences of cooking and cleaning for European Americans for approximately $3 per week. LaShaundra apples]; they won't let you sell 'em. If I'm selling you some, then I don't have to tag 'em but if I'm selling to a store then I have to tag 'em [so] they can tell where they come from. The water has to be tested. He [an inspector] has to come and look in your kitchen, in your drawers, and everything. They've got to know where you get your flour and your milk. Your milk, you have to buy your milk. We had a cow that gives six gallons of milk but we couldn't use it. And, well, we had some chickens too but he said you can't use your chickens, can't use your eggs, have to buy your eggs. So, this cooking is expensive now. It didn't use to be. Like I said, we made many things for ourselves.

"Like my daddy killed hogs. He would go anywhere they had hogs to slaughter. If it was 12 o'clock at night when he got home we would get out of bed, and cleaned them [the hogs] and took the fat off so we could use the fat to make our own soap. You got a box of lye, took all your fat and put it in a sack. The sack went in a big black pot like that one out there and you cooked it. You cooked it outside 'cause it takes it all day. And you cooked it during a new moon. It [the new moon] makes it pretty and white. If you do it when it is a full moon, it rises up out of the sack and out of the fire. There's certain things you can't do in a full moon because it causes it to rise."

She reaches for a 2007 Almanac located on the table at which we are sitting. She arbitrarily turns to March 2007 and begins to explain to me the new and full moon and what you should do and not do during those cycles.

"If you're going to make sauerkraut or pickled beets, you do it on the new moon so the water will stay up over them and it wouldn't take them as long. We didn't know to use vinegar back then when we was making kraut. And that's what took us so long. Now, if I

spoke of a new mode of denied and restricted access: the control of information via the poor advice provided by teachers and the neglect of guidance counselors. This denied and delimited access is accompanied and perpetuated by cultural hegemony, a term I use to refer to the valuing and dominance of one culture over another such that the valued culture becomes the norm. The triple quandary casts mainstream U.S. culture as the valued culture and African-rooted culture as its antithesis.

In brief, a few cluster values distinguish the mainstream U.S. and African-rooted culture. In the U.S. culture, individuals and individual rights are paramount whereas the group and relationships among group members are of greatest importance in the African-rooted culture. Human interactions in a situation are highly valued in the African-rooted culture; in the U.S. mainstream culture, what is tangibly produced from the interactions is key. The U.S. mainstream advocates the development of rational thought and the repression of emotion; in contrast, African-rooted culture nurtures emotional expression as a part of rational thought. To make sense of reality from the standpoint of U.S. mainstream culture, reality is separated and reduced to the observable and material; in the African-rooted culture, reality is perceived as a harmonious whole that extends beyond the realm of the physical into the spiritual. In the life experiences of the two African American females featured in this chapter, LaShaundra's recount of completing school tasks as a family affair and her discomfort with "everyone doing their own thing" clearly exemplifies the hegemony of the was making kraut I would cut my cabbage, put a cup of salt in it and a cup of vinegar. Boil my water and put it over the cabbage. Make sure the water comes up over the cabbage; put a lid on it. [In a few days] I be ready to can it."

As the afternoon faded, Ms. Herlton's voice softened as she spoke, fluctuating between her early years and the not-so-distant past.

"I looked after my grandma. She had cancer. She was operated on twice—that was before they ever started taking Black people into [the local hospital]. So, we lived not too far and I would go give her a bath, comb her hair, feed her, and take care of her. They treated her but they had houses that you could go to down below the railroad. Doctors and 'em would go down with a nurse and change it [bandages], and look at her but when it came time to go home they [her grandmother and other Blacks] had to go home. But they did not, for years and years and years, take Black people into the hospital."

After a few comments about life, stress, and health, her reminiscing focused on the day one of her daughters died a few years earlier from cancer.

"I just prayed to the Lord, 'cause she would come here and stay a lot of times, [that] she wouldn't go by herself. I said, 'Lord, please don't let her die by herself.' So, Grandpa [her husband] and I went to town that morning. When we got back, she [her daughter] had already called [another daughter of Ms. Herlton's], and they were going to go out but she got sick. [A neighbor] motioned me to come and said that [my daughter] had gotten sick. I walked in and I said, 'What's the matter honey?' She said, 'My stomach hurts me so bad.' So I said, 'Turn over. Let's get this soiled sheet out from under you.' And she said, 'Lord, help me. Somebody help me.' I said, 'He will.' So, I got the soiled sheet out and put my hand

mainstream culture and the subsequent tensions that emerge.

Because the low status of African Americans in the US is attributed to racial oppression and cultural hegemony, and African Americans have historically resisted and presently defy this positioning and its determinants, the realms of experience as depicted in the triple quandary are oppositional and incommensurable. The triple quandary and the oppositional and incommensurable relationship among its three components are important in understanding the experiences of African Americans in the contexts in which they develop and function. Cole's CHAT and Uri Bronfenbrenner's ecological environments provide a general view of these contexts.

Cole's CHAT and Bronfenbrenner's Ecological Environment

In Cole's CHAT, culture is central to humans' interactions with their surroundings; to indicate its significance Cole labels individuals' interactions with others and the physical world as cultural practices. Cole further contended that these cultural practices occur over time and are considered a social inheritance that is passed from one generation to the next. The time domains articulated by Cole included (a) phylogenesis, the history of human species; (b) cultural-historical, the history of the cultural group in which an individual is born; (c) ontogeny, the history of an individual; and (d) and microgenesis, the moment-to-moment interactions encompassed in an experience. The cultural-historical, ontogeny, and microgenesis domains are on her. She said, 'Jesus.' And then she was gone. Sometimes people say, 'I wonder if the Lord hears me?' He hears you. God gives me my heart's desires. There's a lot of things I've asked Him for. It was rough coming up but He has always heard my cry."

Ms. Herlton rose and scurried as quickly as she could to the kitchen to check the simmering apples. Her voice livened as she began to dispense words of wisdom to me about marriage, child rearing, and treating other people right.

Recent Past and Present: The Lived Experiences of LaShaundra

"I'm the oldest and the only girl. You have a lot of responsibilities at least in my household. My brothers will even tell you that I'm kind of like the second Mom. I did a lot of making sure everybody is doing their homework, the whole house is clean, cooking, that type of thing. I really didn't think about if it was getting in the way or anything. I just knew it was what I had to do. I thought all older siblings had to do that.

"I went to Christian school from kindergarten to second grade . . . during key developmental periods of my life. Christianity was a part of their education. A lot of my moral values come from that. A lot of who I am is still based upon the fact that I was brought up in a Christian school.

"My brother and I were the only two Black students in the whole school. I noticed that but . . . it didn't feel like it was a racist environment. I noticed that we were the only African American students but I didn't feel 'less than' at that school because of it. Growing up in the Christian school, I didn't really see, but then when we moved to the neighborhood I was really surprised to see it [racism]. [The neighborhood] was mostly White and country. We were

pertinent in my re-examination of authenticity. In addition to viewing cultural practices with respect to time, Cole also asserted that these practices must be examined within the institutional arrangements in which they occur. Bronfenbrenner's ecological environments are one characterization of these institutional arrangements.

Bronfenbrenner posited that an individual's direct and indirect connections to and interactions with settings and conditions within those settings comprise the individual's ecological environment. An individual's ecological environments are represented by a set of nested, concentric circles that are treated as a system. The innermost circle is called the microsystem and encompasses the most immediate settings in which the individual is directly involved. These settings are marked by a defining set of material and symbolic characteristics and the individual's state within them is greatly influenced by the pattern of activities, roles, and interpersonal relations the individual perceives and experiences. The mesosystem, the next level, represents the interconnections among two or more microsystems; that is, the mesosystem captures the interrelations among the various settings in which the individual participates. The links between the microsystems and the various forms the links may take (e.g., interactions, expectations) constitute the mesosystem. Unlike the microsystems and corresponding mesosystems, the individual is not directly involved in the third level, the exosystem. The exosystem represents institutional arrangements (e.g., school district administrative offices) that are affected by or affect the one of the few Black families in the neighborhood. There were other Black people but they were older, they were older kids so they really didn't hang out with us. The only kids our age were White. And we were one of the first Black families in that neighborhood with kids that were young and them getting used to us. Typically, you wouldn't have a problem but there were certain homes that we couldn't hang out at. There were certain kids that we could only play with them outside; we could never go to their house or anything. The neighborhood with the White kids, that really impacted me because those racial issues—when you're a kid and you just want to play. I think it impacts the way I look at education and look at racial issues in general to this day.

"[Going from Christian school to public school] I did start to notice that teacher expectations were different for me as a Black student. Before, no one ever told me that I wasn't smarter; teachers didn't seem to have that low expectation. My parents, my mom is a social worker and my dad is a counselor, were really involved in Christian school and they could come by and talk to the teachers. All of a sudden in public schools, it was totally different. It was almost an antagonistic relationship between teachers and parents. Even in third grade, I understood that I had to learn a different way of negotiating school because there were rules in school that I didn't necessarily know—from riding the bus, to where the Black kids sat versus the White kids, to who your friends were supposed to be, to simply how to line up for lunch. [I had to learn] a whole different culture. I was the only one [Black] in my classes—not saying that it didn't bother me because it did bother me. I had to deal with a lot of assumptions from the White kids and the White teachers.

settings in which the individual participates but these structures do not involve the individual as an active participant. The microsystems, mesosystems, and exosystems are encased within the last level, the macrosystem. Patterns that exist at the level of a subculture or culture and the beliefs and ideologies that underpin these consistencies constitute the macrosystem (e.g., societies).

Each model—the triple quandary, CHAT, and the ecological system—in isolation is insufficient in examining the development and experiences of African Americans as a collective. To gain a more comprehensive understanding of phenomena with respect to the African American experience in the US, a phrase that captures the commonalities among individuals who are classified as African American, I fused the three models to form the TQS CHAT-Ecological framework (see Figure 2.1).

The Synthesis

In Figure 2.1, the CHAT components—cultural-historical time microgenesis, and ontogeny—are in the background. The micro-, meso-, exo-, and macro-aspects of the system of ecological environments, as nested circles are in the foreground. The triple quandary, represented as lightning bolts, is superimposed upon CHAT and the system of ecological environments. In gaining a comprehensive understanding of the experiences of individuals separate from and a part of a group, the individual or the collective is placed at the core of the framework. In this reconsideration of authenticity, I place African Americans in the centre of the microsystem ("micro" in Figure 2.1).

"When I was in the junior or senior year of high school, we had to write a paper and I wrote mine on *Uncle Tom's Cabin*. I really put my heart and soul into that paper. It was a huge paper and it made up a tremendous amount of our grade. I got it back and I made a '40.' Of course, I'm devastated because I put so much into it. And what she [the English teacher] had done was to take off five points for every typing mistake I had made. My mom typed the paper for me because we did not have a computer at home. So we had to go to my mom's job where she would have to work after work and some times we would be there for hours. My mom typed one space between sentences but there were supposed to be two spaces. What my English teacher did was to go through and for every place there was one space she took off five points so I failed that paper. And she didn't say one word about the substance of the paper. It did not reflect my effort and didn't reflect what I knew but that's what I got because of typing errors. It really hit home with me what that grade was reflecting [was] really her own personal bias.

"And another class would be in my geometry class. We had to do a geometry project. The first thing you had to turn in was a small one like on a mini-scale. It was like a science fair but it was geometry. So I did mine on fractals. I had made this small little board out of construction paper and put small diagrams of fractals over the board. I got a 95 on it. What I did was make that small board into a giant board because that was the point of the small board [which] was to show her what we were going to do; I turned that in and she failed me. She said that she misunderstood what I had done on the small board and that I did not properly convey it on the big board. Still today, I

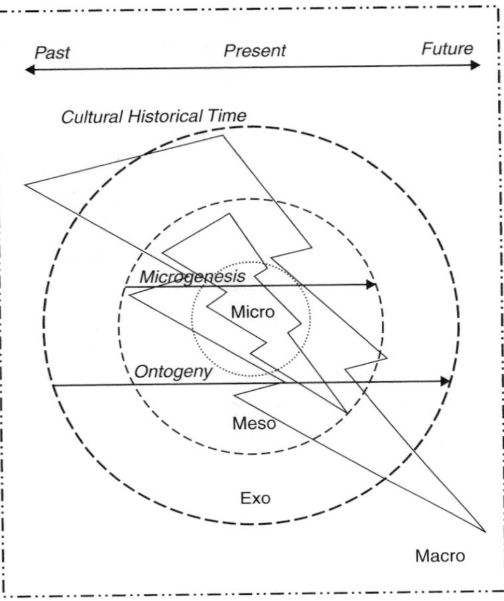

Figure 2.1 The components of the ecological system are micro, meso, exo, and macro positioned within the CHAT time-frames of cultural historical, microgenesis, and ontogeny. The lightning bolts represent the triple quandary; the small one is the African-rooted culture realm and the large one encompasses oppression and mainstream culture domains.

As indicated in Figure 2.1, what transpires within the ecological environments (micro, meso, exo, macro) occurs within broader dimensions in the form of cultural-historical time. The use of cultural-historical time necessitates a consideration of events in relation to the group's past, present, and future. For example, in this chapter that features the life stories of two African American women, Ms. Herlton's childhood, adolescent, and young adult years living under apartheid in the US and LaShaundra's upbringing in a desegregated society are relevant and significant. Time domains more proximal to individuals than don't really understand what she meant but I failed that project. Again, to me, it was a case of not what I understood, not my real effort.

"I had a great band teacher. And out of all the teachers that I had he was really influential in my life because he was one of the first teachers and this was in high school in the tenth grade that really encouraged me and really thought that I had some kind of potential. I never really had in public school a teacher like that in my life. I always felt like I had to prove myself and even when I did, the expectations weren't really that great. Teachers would be surprised if I did well. He was the first teacher, he wasn't from [resident's state] but he was from Pennsylvania; that would tell me 'you have a lot of potential,' 'you can do this.' Whenever I was around him, I always felt like I could actually do what he was expecting. I could tell a huge difference in my self-esteem after having him as a teacher. Sometimes I still send him Christmas cards and everything just to let him know that really did impact my life.

"One of the things that really bothers me sometimes about school: I think kids are expected to do a lot of things that [the] middle class or upper middle class have privileges to but [they are] not . . . in those classes. For instance, reports. My parents took me to the library because they were knowledgeable. There were a lot of reports I had to do [and] my parents could direct me to the right resources. Students don't necessarily have the resources, access, or knowledge to go to those sources for, say, to do a leaf project. Well, my father thought, 'Well, I'll take her to a plant store where they actually have the names of the plants there with the leaves.' If he hadn't thought to do that I don't know what I would have done if I was left to my own devices. Or just the

cultural-historical are also noted in the framework.

Microgenesis, experiences in the present, is delimited to the micro and meso, ecological systems that encompass individuals' most immediate settings in which they directly participate. Ms. Herlton's baking apple pies within her home is an example of microgenesis within the ecological microsystem. Her account of tagging baked goods made with home-produced ingredients (e.g., milk from her cow) to be sold to individual buyers versus baked goods made with store-bought ingredients to be purchased by retailers exemplifies microgenesis within the mesosystem. In the previous example, the microsystems surrounding her personal contacts, like friends or community members, and her business contacts, like the local store, constitute the mesosystem.

Last, ontogeny, the history of entities of interest, cuts across the surroundings in which individuals directly participate and extends into the exo- and macrosystems, life-impacting ecological environments somewhat removed from them. LaShaundra's journey to a degree in meteorology is an exemplar of ontogeny. When she described the support of her family, she highlighted the microsystem of the home. The tensions (e.g., new cultural codes to decipher, antagonistic relationship between parents and teachers) that existed in her transition from a Christian school to public schools are exemplars of the mesosystem (i.e., the relationship between the microsystems of home and school). Her comments regarding the mismatch between demands in the classroom (e.g., "kids are expected to do a lot of things that [the] middle class or upper middle class have privileges fact [that] science is not my parents' area; it's just not. I just happened to have that love. The expectations of other people for me going through a meteorology degree were not very high. The science teachers did not encourage me for that. In fact, they gave me bad advice. There were no role models. I never had a Black science teacher. Actually, I only had one Black teacher in school prior to college and he was my history teacher. But, he had issues. If my parents hadn't encouraged me. I wouldn't have had enough math to get into [university for undergraduate study]. They had to be advocates for me. My guidance counselor wasn't looking out for me in that regard. Also, just experiences. Like from [a small rural town] there's not a lot of museums, not a lot of places I could have gone to get that interest [meteorology] in me driving then because it wasn't being met in school. I always wished that we would talk about some kind of weather in class and we never did. That learning would just come from daily weather; we had cable so I could watch the weather channel. Other than that, it [science interest] wasn't fulfilled in any other kind of way. That need was not being met except on my own, which was looking at encyclopedias, reading stuff from the library. So I don't know what would have happened if my parents hadn't known about those resources."

With the numerous challenges she faced in school, I wondered how she persisted to secure science and graduate degrees. As she moved from discussing the challenges in school, she lowered the intensity and loudness of her voice.

"I was in Christian school . . . on the playground and I happened to overhear two teachers talking about this thing that comes in the night and it sounds like a freight train. It grabbed my imagination. At that time, I thought it must sound like to . . .") and the resources available to students (e.g., "My parents took me to the library because they were knowledgeable") implicate the exo- and macrosystems.

In the synthesized framework, the examination of experiences or events in relation to African Americans within the CHAT time domains and Bronfenbrenner's ecological environments are also concurrently viewed in regard to the group's distinctive predicament—the triple quandary.

To signify the immersion of individuals (i.e., African Americans) in cultural practices most germane to them, the African-rooted culture realm of the triple quandary is positioned within microgenesis and the corresponding ecologies of the micro and meso. African-rooted cultural practices are evident in the micro dimensions of Ms. Herlton's and LaShaundra's lives. One example is the emphasis of spirituality by both Ms. Herlton and LaShaundra. Because the remaining two aspects, oppression and mainstream culture, are situated in the quandary as hegemonic, the figure depicts them as crossing all time and ecological domains.

The engulfing and enclosure of the African-rooted culture within the U.S. mainstream culture in the figure is an attempt to depict the contentious relation of the two culture-related realms of the triple quandary. Ms. Herlton and LaShaundra shared numerous examples of oppression and the dominance of mainstream culture over African-rooted culture. Among them, Ms. Herlton spoke about the segregation of African Americans in receiving medical services. LaShaundra described the personal conflicts that emerged from the a choo-choo train. So I imagined something coming in the night going 'choo-choo.' Once I heard them say it was a tornado, I went to the library and read as many books on it that I could. I read one particular book that spoke about how tornados are formed but had stories of people who had lived through them and I was hooked. Everything I could get my hands on I read; every time they had a special on TV, I'd watch it. When I was a little kid, every thunderstorm that came . . . my dad, we wouldn't go inside the house; we would sit on the porch until it went by. I would just love that. It was just really neat. It [love of weather] has always been a part of who I am. Right now, I will go to friends' homes and do some kind of weather activity with their kids. I went to an elementary school [recently] and presented weather information to their fifth graders. I don't think I'll ever lose that love for the weather.

"For me, it's like an instinct to go and always check the weather. Not just the local weather. I'll also go to the weather channel to look at the United States to see what is the weather doing to the west so I can see what I can expect later on. Or, say that it's raining. I will look at the clouds and I will try to figure out which way the clouds are going to find out what kind of weather system we're having. For instance, I can tell that if the rain is coming from the south then I'm pretty sure we have some kind of disturbance on the coast instead of just the normal weather pattern, which comes from the west to the east. From that moment on [the playground], I decided that I wanted to study tornadoes. Later on, I found out that it was meteorology. So, that's what I knew I wanted to go to school to do.

"That love for the weather is really what got me through school and my degrees the whole time. And also my mainstream cultural value of individualism and the African-rooted cultural value of communalism in the context of completing pre-college and graduate school assignments. This re-conceptualization of the African American experience, TQS CHAT-Ecological approach, requires re-thinking phenomena and constructs often examined in relation to African Americans. I revisit and reconsider authenticity as it is currently employed in the science education literature.

Authenticity: A Second Look

In Buxton's (2006) synopsis regarding what is the meaning of authenticity, the micro-level in the TQS CHAT-Ecological framework is addressed in the canonical, youth-centered, and contextual classifications. For example, the canonical view defined authenticity via the micro-world attributed to scientists; and the youth-centered and contextual perspectives positioned authenticity in the micro-worlds of youth. With respect to the desired ends of authenticity, the ecological levels emphasized in the literature grouped in the youth-centered classification differed from the level highlighted in the canonical and contextual perspectives. The canonical and contextual views situated outcomes in the macro-level (e.g., a scientifically literate populace in terms of scientific knowledge and the nature of science to contribute to society). In contrast, the goals of the youth-centered view of authenticity addressed all levels in the ecological environment system (e.g., to solve real-life problems in students' lives and society). Considering what authenticity means in its current portrayal in the science education family telling me that I could do it. Growing up, whenever someone in the family had something to do, the family had something to do. It wasn't like an individual—like this is your project and you have to go upstairs and work on it. The whole family knew what you had to do and the whole family was involved. If that was helping you to think about it or come up with ideas or thinking up resources. If you had a leaf project then everybody would go to the park and get some leaves with you. Every project was a family thing; it was never an individual thing. I get surprised when I go to someone's home and it's like 'He has homework' and everybody else is doing their own thing. Even in my marriage, when I was doing my dissertation, first my husband [who is not African American] was like 'That's your thing'; towards the end he read over my work because I missed that 'whole communal everybody's involved' piece. That individual thing, it seems so cold. It's not part of who I am.

"I knew what I had to do to be able to do what I loved. Even though [others] may not have high expectations for you, you still got to do the best that you can do. When there were days when I did not want to go to meteorology classes, I would think of people in the past, for instance slaves, that were told that they could not [learn to] read or write, or [face] punishment or penalty of death. Just my grandparents who worked in factories, that [they] just didn't have the luxuries that I have, so I would make myself get up and do those things. I guess the love of the weather [and] an expectation to do things, not just for me but for my family and African Americans as a whole. It was bigger than me. Everything was bigger than me. This community effort to get me through school and I was thinking of the community when I got out."

literature does not consider the distinctive conditions of African Americans. The previous statement is illustrated in a closer examination of Buxton's review in relation to two questions: "What does authenticity mean according to whom?" and "Authenticity for whom and for what purposes?"

Numerous responses to the previously stated questions are implied in literature. The canonical view, with its definition of authenticity residing in the micro-worlds of scientists (who are predominantly European American) within scientific communities (which are governed by codes of those who dominate in society) promotes what exists and does not forge connections with the individuals deemed in need of authenticity in science education. The youth-centered approach values and utilizes the micro-worlds of youth but fails to acknowledge the implications of utilizing these micro-worlds for negotiating the meso and accessing the exo and macro. Limitations similar to those for the youth-centered approach exist for the contextual view, except the contextual view considers access to the exo and macro levels of the ecological environment system. If I answer "What does authenticity mean according to whom?" and "Authenticity for whom and for what purposes?" from the standpoint of the African American condition in the US as it is framed by the TQS CHAT-Ecological framework, then what would authenticity look like? Tempered by my perspective as an African American female whose life spaces cross the positions presented in the two parallel presentations, I reconsider authenticity by merging TQS CHAT-Ecological framework and Ms. Herlton's and LaShaundra's lived experiences.

Taking a Stand(point): Authenticity in Science Education for African Americans

As indicated in the TQS CHAT-Ecological Model, authenticity for African Americans must be multidimensional and all dimensions must operate in tandem at numerous levels. Authenticity must be simultaneously *particular* to the perspectives and lives of African Americans, *preparatory* for accessing and then transforming the existing establishment, and *political* in goals to improve the positioning of African Americans in the US. I denote the pertinent aspects of the TQS CHAT-Ecological Model in parentheses.

The particularities of an authenticity in science education for African Americans must respectfully incorporate the perspectives and life experiences of African Americans (microgenesis in the micro and meso; ontogeny in the meso; and cultural-historical in the macro). For example, even though Ms. Herlton did not know the canonical name of *Lobelia inflata* and could not describe how it remedied bronchitis and arthritis, she knew in what forms (i.e., dried, liquid extract, tinctures) to use *Lobelia* and for which ailments. To be authentic for African Americans, the particularities of authenticity cannot be restricted to knowledge but they must also address environing circumstances as perceived by African Americans (oppression of TQS; microgenesis in the micro and meso; ontogeny in the meso; and cultural-historical in the macro).

LaShaundra provided example after example of undercurrents that she perceived to work against her in her pursuit of becoming a meteorologist. The forces that countered her efforts included low teacher expectations, unjustifiable and unsubstantiated evaluations that LaShaundra attributed to personal bias, and misguided, erroneous, or withheld advice on the part of school personnel regarding her academic endeavors. An authenticity in science education for African Americans promotes a critical consciousness among authorities working with African Americans that leads to the acknowledgment, curtailment, and ultimate elimination of these informal and unjust practices (see also Emdin, chapter 5, this volume). In addition to valuing and addressing the particulars of African American perspectives and experiences, the particularities must be utilized in a manner that engenders an authenticity that is also preparatory.

When the knowledge from or related to the micro-worlds of African Americans is the origin of authenticity, the need to value and respect African Americans' funds of knowledge should be tempered by the implications for access to knowledge that is presently privileged in the US (microgenesis in micro and meso; ontogeny in exo; cultural-historical in macro; cultural hegemony in TQS). That is, authenticity in science education for African Americans not only affirms their perspectives and experiences but it also prepares them to access opportunities beyond their micro-worlds. For example, Ms. Herlton described what one should do and not do during a full moon or new moon. She spoke of the influence of the moon cycles upon soap making and the canning of vegetables and fruits. Canonical science, privileged knowledge in the US, declares the impact of moon cycles upon human events as folklore or a myth. In authenticity for preparation, the divergent perspectives—Ms. Herlton's and that of canonical science—and their underlying premises are explicated and explored with the goals of culturally sensitive criticality and competence as outcomes. Like in the case of particularities, authenticity for preparation extends beyond competence in content.

Authenticity in science education that is preparatory for African Americans must equip them to negotiate the tensions among and within their ecological environments and to maintain a positive sense of self while at the same time challenging and transforming oppressive and unjust conditions (microgenesis in micro and meso; ontogeny in micro, meso, and exo; cultural-historical in micro, meso, exo, and macro; oppression and cultural hegemony in TQS). LaShaundra's experiences of learning how to navigate school and its cultural norms as young as third grade, and how to deal with racial issues (e.g., assumptions of White students and White teachers), are testaments of this previously described authenticity. Because she learned how to navigate the worlds of home, community, and school, she was able to succeed within the school environment and develop a positive sense of self, one that was independent of and connected to African Americans as a collective. Authenticity in science education that is preparatory for African Americans cannot exist without authenticity that is political in its aims.

In light of the low status of African Americans in the social, cultural, educational, economic, and political spheres of American life, seriously examining power and scrutinizing the mechanisms through which certain groups impose their beliefs and practices upon others is imperative for any authenticity in science education for African Americans. Authenticity that is political in its goals must disrupt hegemonic institutional structures and ideologies. With the intent to eradicate the low-stationing of African Americans in U.S. society, authenticity in science education that is political in nature reveals and openly situates the contemporary experiences (microgenesis and ontogeny) of African Americans at the micro-, meso-, exo-, and macro-ecological levels in relation to the cultural-historical, the history of African Americans as a cultural group in the US. When taken together, the life stories of Ms. Herlton and LaShaundra illustrate the connection between the contemporary and the cultural-historical. Ms. Herlton's life space clearly demonstrated the impact of legalized racial oppression and cultural hegemony and LaShaundra's life sphere showed progress riddled with covert remainders from the past.

Coda

In its current state, authenticity in science education falls short of being authentic for African Americans. Moving beyond the rhetoric of science education reform requires a re-visitation, reconsideration, and rethinking that is taxing, unsettling, and unpopular. Perhaps this merging from the standpoint of an African American female of the life spaces, separated by a half century, of two African American women with the TQS CHAT-Ecological Model provides a viable locus from which to proceed.

References

Boykin, A. W. (1986). The triple quandary and the schooling of Afro-American children. In U. Neisser (Ed.), *The school achievement of minority children* (pp. 57–92). Hillsdale, NJ: Lawrence Erlbaum Associates.

Boykin, A. W. (1994). Harvesting talent and culture: African American children and educational reform. In R. J. Rossi (Ed.), *Schools and students at risk: Context and framework for positive change* (pp. 116–138). New York: Teachers College Press.

Bronfenbrenner, U. (1979). *The ecology of human development: Experiments by nature and design.* Cambridge, MA: Harvard University Press.

Buxton, C. (2006). Creating contextually authentic science in a "low-performing" urban elementary school. *Journal of Research in Science Teaching, 43*, 695–721.

Cole, M. (1996). *Cultural psychology: A once and future discipline.* Cambridge, MA: Harvard University Press.

Parsons, E. C. (2008). Positionality of African Americans and a theoretical accommodation of it: Re-thinking science education research. *Science Education, 92*, 1127–1144.

3 Faith in a Seed

Social Memory, Local Knowledge, and Scientific Practice

Carol B. Brandt

> You plant these bean seeds when the oak leaves appear, when the red hummingbird comes from the south. You can plant and know the ground will be warm. The hummingbird never lies.
>
> (Anselmo Quam, 1991)

Above my desk is a photo of rough, calloused hands cradling the large bean seeds—vivid pink and streaked with maroon. The photograph (Figure 3.1) was taken as Anselmo[1] told me of the ways in which he gauged the proper time for planting. Anselmo's knowledge was vibrant, alive, and relevant to his community at the Pueblo of Zuni. For me, this photo is a reminder of a knowledge system that connects people in meaningful ways to their everyday world.

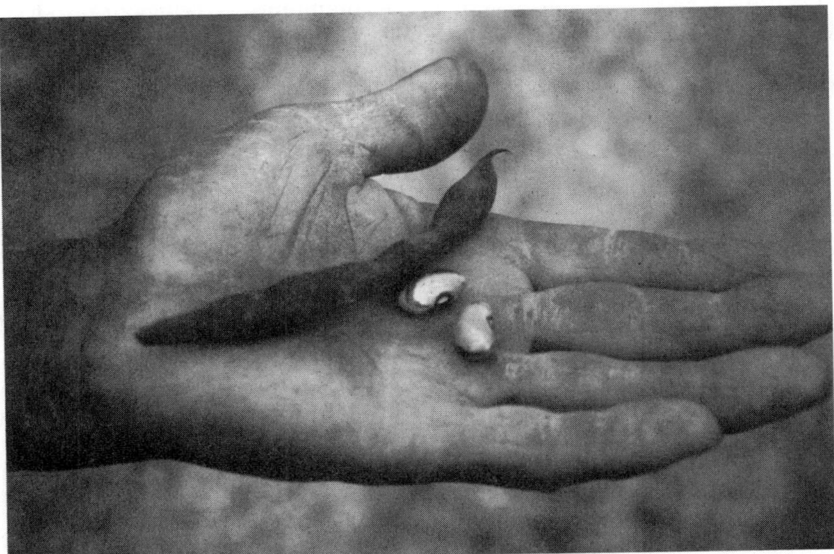

Figure 3.1 Seeds of a Zuni variety of string bean grown by Anselmo.

My research on seed-saving strategies, conducted with traditional farmers and gardeners like Anselmo, inspired me to rethink my own views of teaching science in the classroom. After working for three years with Zuni farmers, I was able to see a complex agro-ecological system. As part of a study for the Pueblo of Zuni, farmers discussed with me genetics, soil morphology, hydrology, and botany, all using their own terms with the sophistication and confidence. This experience at the Pueblo of Zuni resonated deeply with my own love of gardening and my first experiences of inquiry in the natural world through agriculture in the Midwestern US. Speaking with Anselmo and other farmers made me wonder: Why do we teach science so disconnected from the daily lives of our students? What are the dangers of teaching science as an objective, an invulnerable truth, with few connections to students' lived experience?

In the six years that I lived at the Pueblo of Zuni, I glimpsed a vast cultural geography, as well as a new definition of landscape and its connection to traditional knowledge. My Zuni co-workers taught me to recognize reeds, cattails, and tadpoles as keepers of rain. Farmers introduced me to new meanings for seeds and their connection to rebirth and cultural continuity. I began to see geological features along the Zuni River, not in terms of their geological history, but in terms of the Zuni myths and seasonal passages. These understandings made me wonder at the relevance of my scientific practice and inspired me to ask: where is the living body of science?

In this chapter, I explore the everyday world of gardening and farming, both from my own experience growing up in the Midwest and in light of my research at the Pueblo of Zuni. Much like formal schooling, sites for learning science outside of schools are framed and shaped by social, economic, and political discourses. Being raised in an agricultural community, growing our own food was second nature to daily life in my small town. Gardening was woven into the "common sense" of our community and along with other youth, I attended 4-H meetings organized by the county agricultural extension agents where I began to record "scientific data" on yields and to experiment with new varieties. Agriculture in my German-American community was infused with Christian values and a staunch faith in Eurocentric science as a result of the Progressive Era that revolutionized Midwestern farming in the 1920s. Attached to this local knowledge and informal science were our ways of talking about time and space that also ordered community practices.

None of my experiences, however, prepared me for gardening in west-central New Mexico. On the high desert plateau, this arid environment is marginal for growing conventional crops and yet, Zuni is the home to one of the oldest Indigenous agricultural traditions in the United States. When I began working with Zuni farmers and gardeners my notions of "common sense" in agricultural practices shifted when I examined gardening as the relationship among local knowledge, economics, and the political history of Eurocentric science in this Indigenous community. In this chapter, I chronicle how I grasped the role of social memory in maintaining local knowledge in an agricultural repertoire despite dramatic economic and social changes. In learning how to garden at

Zuni, I came to understand how Eurocentric science is part of a larger custodial discourse between the federal government and Indigenous people. Drawing from interviews with 50 gardeners and farmers at Zuni, I describe ways in which local knowledge is parsed and dispersed in the community, and how local conceptions of space-time contrasts with Eurocentric science.

Like Dorothy Smith (1996), I use the metaphor of a map to show how knowledge as a social product reflects the relationship among various local sites of experience—without undermining any one of these sites. The map enables us to locate ourselves within the complex relationships of others in which we are embedded and act. For Smith, the purpose of sociological inquiry is to conduct research *for* the people and raising our consciousness of daily life, rather than one that is *about* people. This approach starts with one's experience, but she does not equate experience with knowledge. Instead, Smith uses experience as a place to begin an inquiry into daily life, to see how an individual's consciousness is situated in larger on-going social relations and organizational structures.

> We begin with a knower, a subject whose everyday world is determined, shaped, organized by social processes beyond her experience and arising out of the interrelations of many such experienced worlds. They are relations that coordinate and codetermine the worlds, activities, and experiences of people entered into them at different points.
>
> (Smith, 1996, p. 134)

A New Cultural Cartography

As I crossed the continental divide, my small truck was enveloped in a swirl of snow. A late spring snowstorm had swept down from the Rockies and I was driving to my new job as an archaeologist at the Pueblo of Zuni, on the western border of New Mexico and Arizona. The storm was one of those disorienting blizzards where the fierce snowfall and shifting wind made me feel as if I were going backwards, to the right and then forwards, even as I gripped the steering wheel and aimed for the stripes that bordered the highway. As the snow and clouds parted, I caught a glimpse of steep canyon walls and the shadows of sandstone pillars erupting from an endless sea of scrub pine and juniper. In the snowstorm, I wondered, what are my points of reference? How do I navigate this new landscape?

With a new MSc in botany in hand and years of experience as an archaeologist, I had accepted a position at the Zuni Cultural Resource Program where I would be working in their archaeology laboratory analyzing wood, seeds, and other plant materials from archaeological sites. Little did I know that the disorientation I experienced in that snowstorm would later become a metaphor for my work and life at Zuni. Gradually, I realized how my points of reference and use of Eurocentric science in this new cultural landscape were of limited assistance. What started as a month-long position evolved into a stay that lasted for nearly six years, during which I came to think in new ways about the relationship among Eurocentric science, agriculture, plants, and people.

Zuni—The Middle Place

Among anthropologists, Zuni is legendary. Isolated in the mesas of western New Mexico, the pueblo and its Indigenous community has attracted anthropologists, linguists, and archaeologists for more than a hundred years. The pueblo's elaborate religious ceremonies, prehistoric ruins, and art captured the imagination of eastern intellectuals. Zuni is often cited in American history texts as the location where the Spanish explorer, Francisco Coronado, clashed with the Zuni people at the village of Hawikuh in the early 16th century. Coronado and his motley entourage encountered the Zuni people in their search for the golden seven cities of Cibola. Beyond their claim to a role in early Spanish history, Zuni farmers were renowned for establishing a complex agricultural economy adapted over hundreds of generations in this high, dry elevation of the Southwestern US.

Like many academics before me, I was mesmerized by romantic representations of the Zuni Pueblo. My imagination had been fueled by the writings of 19th century ethnographers Frank Hamilton Cushing and Mathilda Coxe Stevenson. Both wrote extensively about Zuni plant lore and agriculture. These ethnographers came to Zuni during the late 1880s employed by the Bureau of American Ethnology (BAE), a research arm of the United States Department of Interior. Their expeditions to Zuni were part of the Bureau's mission to document Indigenous culture in the rapidly changing western states. Bluntly, the BAE was predicting the extinction of Indigenous cultures in a time when the U.S. government was bent on a program of forced assimilation.

The Zuni Indian Reservation is located in west-central New Mexico, between the Arizona–New Mexico border and the Continental Divide, encompassing roughly 640 square miles (Figure 3.2). The elevation ranges from 6,031 ft (1,838 m) along the Zuni River in the southwestern portion of the reservation to 7,000 ft (2,134 m) in the northeast corner near the Zuni Mountains. The *A:shiwi*, the Zuni people, have endured at their "middle place," *Halona:itiwana*, for more than a thousand years. According to Zuni oral history, their ancestors wandered the Southwest for generations until they found their homeland, the middle place. Their continued existence in this high, arid environment is in part due to crops, a mixture of maize, beans, squash, sunflowers, and other garden vegetables. However, over the last 150 years, the Zuni tribe has been confronted with a reduction of their traditional lands, the destruction of valuable farmland, and the introduction of a cash economy.

As traditional farming has decreased, individuals maintaining agricultural skills and traditional knowledge, and traditional crop varieties adapted to the high desert elevation of Zuni have become scarce. When tribal lands were circumscribed by the federal government in the late 19th and early 20th centuries, Zuni farmers were not only restricted from traditional farming areas beyond the reservation boundary, but they also found themselves competing with large sheep and cattle herds. Also, degradation of tribal lands occurred from overgrazing by livestock upstream, off the reservation, due to mismanagement of natural resources by the federal government in the early to middle part

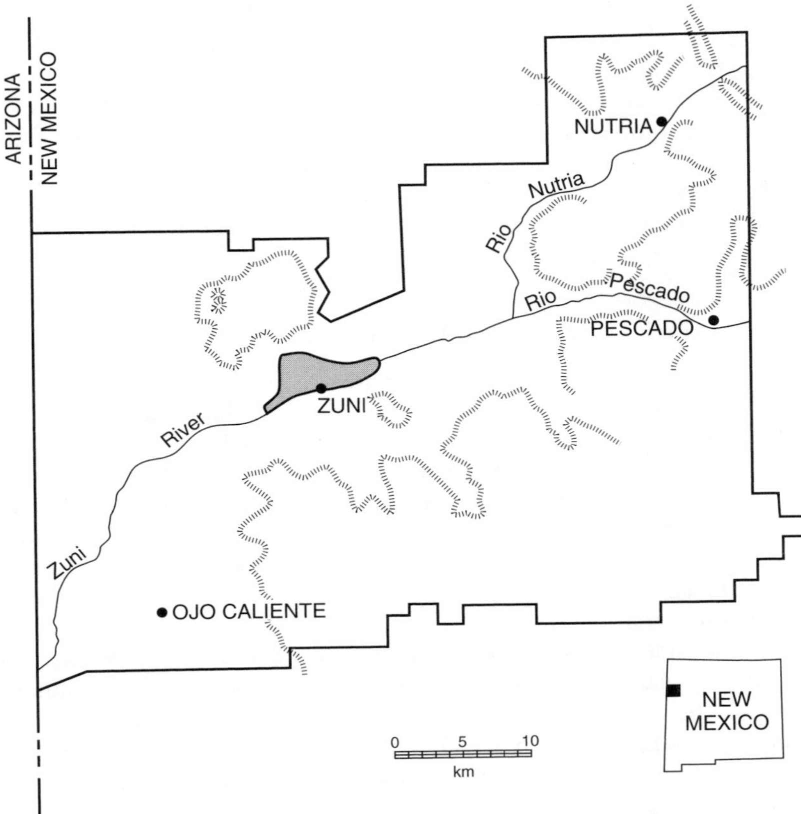

Figure 3.2 Location of the Zuni Indian Reservation and the Pueblo of Zuni.

of the 20th century. Widespread soil erosion from overgrazing, timber cutting, and water control projects resulted in the destruction of valuable farmland.

One of the most notable features of the reservation is the Zuni River, which flows from the east to the west, gradually curving south to the Arizona border where it drains into the Little Colorado River. Run-off from the snowmelt of the Zuni Mountains during the spring provides an important source of water for the Zuni River. Another source of water on the reservation are the springs at Ojo Caliente, Nutria, and Pescado, where ground water is discharged to the surface through faults and other geological structures. Most of Zuni agriculture is dry-land farming, where rainfall is harvested from the surrounding landscape. Fields are strategically placed at the confluence of arroyos and in locations where surface run-off from rainfall events will direct water onto the fields.

The two most critical constraints on agriculture on the Zuni Reservation are precipitation and temperature. Precipitation occurs on the reservation in a distinct seasonal pattern, with late summer rains and winter snowfall. The highest amount of precipitation occurs in the months of July, August, and September in the form of late afternoon or evening thundershowers. This rainfall can be concentrated in torrential downpours. Unless the run-off from these showers is controlled, it can cause soil erosion. The snowfall that occurs from November to March contributes important amounts of moisture that are critical for spring planting. Although the higher elevations have higher levels of rainfall, the distribution of rain and snow can be erratic from year to year over the entire reservation. The high elevation of the reservation gives rise to a very short growing season, limited to 90–100 days.

A Survey of Traditional Zuni Crop Diversity

In my second year at Zuni, I established a garden in the red clay with disastrous results. I had never failed so miserably at gardening, a lifelong interest of mine. All of my previous experience with gardens had developed from the rich loamy earth of central Illinois, but here at Zuni, my stunted plants either succumbed to the unrelenting springtime wind, the searing sun, or the alkaline soil. Seeing small pockets of productive gardens at the springs around Zuni and the many dry-land farms, I realized that I had to completely re-adjust my understanding of gardening and think differently about the relationship between plants and local environmental conditions.

About the same time as my gardening fiasco, the Zuni tribe was developing a new land use plan and little information was available on current agricultural crop resources. With a small grant from the Charles A. and Anne Morrow Lindbergh Foundation, I spent one summer surveying Zuni households on the variety of crops they grew and their seed-saving techniques. By driving out into the Zuni countryside, my colleague, Jerome Zunie and I located Zuni farmers working in their fields. With these farmers, we used the “snowball” method, asking for referrals of other gardeners and farmers to interview as well. In most of the interviews, we went to the gardens and fields to photograph and collect samples of the traditional Zuni cultivars. In all, we interviewed 59 farmers and gardeners in 50 households.

Among the 50 households, a total of 46 men and 13 women participated in our interviews. Nearly half of the participants were over the age of 60 and few of the survey participants were younger than 30 years old. In our interview we also asked how participants made a living; many were commuting to the nearby city of Gallup for wage-earning jobs, a third of our sample was retired, and others supported themselves through cattle or sheep herding and jewelry making. Only two people in our survey identified themselves as “farmers.”

In our interviews we asked farmers and gardeners to name the types of cultivated plants they grew each year in their fields and gardens. We relied on the participants to identify whether they considered their seed to be traditional Zuni varieties or non-Zuni. The most significant feature of our interviews was learning that farmers and gardeners chose Zuni varieties because of their proven success in coping with a short growing season and low rainfall. The most common type of crop grown was Zuni blue maize and white maize, followed by beans, squash, and assorted vegetables. Nearly all of the farmers and gardeners were avid seed-savers, with seed being saved from year to year, purchased from other tribes such as the Hopi or Santo Domingo Pueblo, or from the Night Dances. When people were growing vegetables in an irrigated garden, then they tended to use commercial seed purchased from a seed catalogue. For example, among the blue and white corn varieties (Figure 3.3), most of the maize grown was from locally selected and personally saved seed and grown in dry-land fields.

My colleague Jerome explained to me that seed from the Night Dances is particularly symbolic, as the seed represents blessings distributed by the Zuni ceremonial masked dancers, the Katchina. At ceremonies held in the winter months, attendees bring a handful of seeds to contribute to a basket on the altar. These seeds are then blessed and redistributed to members of each kiva group, a religious community in the pueblo. To realize the blessing inherent in the seed, one must grow out the seed the following spring. We surmised that this custom might be one way to ensure that new genetic diversity is infused into traditional seed stock. Nowadays, the seed from the Night Dances is a mixture from commercial sources and seed saved from previous years.

Like the geographical features on the Zuni landscape, seeds carried by ceremonial dancers at Zuni have a special significance. "Every masked dancer carries a package of seeds in his belt. It is his 'heart.' At the close of any dance the priest who thanks the dancers takes some of the seeds to plant" (Bunzel,

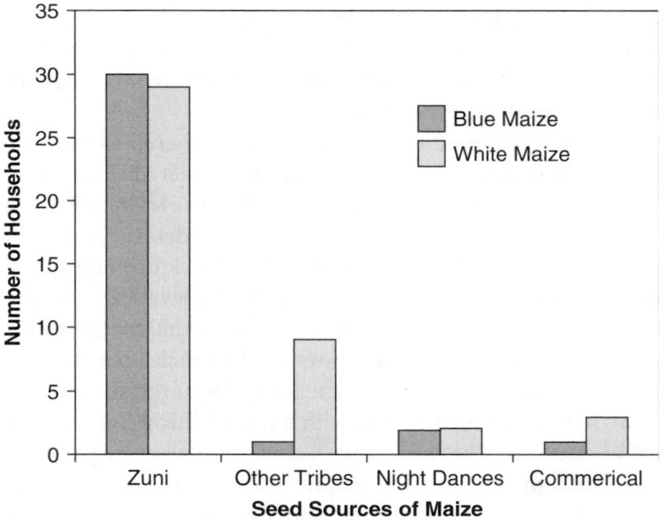

Figure 3.3 Seed sources for blue and white maize varieties among 50 Zuni households, all grown in dry-land fields.

1932, p. 714). In Bunzel's translation of Zuni ritual poetry, Sayataca's Night Chant, the dancer names 57 different seeds in a poem. These seeds are tied into a pouch at his belt: the varieties of maize and beans, wetland plants, grasses, cacti, and shrubs on the mesas. Each seed was deliberately chosen to record and remind the community of their long-held relationship with the landscape, and as a way of honouring the ancestors of the Zuni Pueblo.

As we completed the survey, Jerome and I noticed some unsettling trends. First, we saw little interest in farming among young people in the pueblo; farming no longer seemed to be an economically viable option. Elders who were retired maintained most of the traditional Zuni varieties. Also, no single person had knowledge of the complete agricultural repertoire that was described in the historic accounts. Instead, we found that individuals tended to specialize in a particular crop. For example, Anselmo was an expert in Zuni bean varieties and grew more than a half dozen types of dry-land beans. One woman considered herself to be the guardian of the full range of Zuni maize varieties that represent the six different kiva groups in the pueblo. Another Zuni man, a celebrated Zuni jewelry maker, raised a specific variety of Zuni white corn that was used in Zuni ceremonies. We worried what the future held for the continuation of this biological diversity and the knowledge surrounding Zuni farming practices.

Deciphering the Agricultural Landscape

As we completed the survey of Zuni traditional crop diversity, Jerome and I drove our battered truck to the far reaches of the reservation and I began to see the landscape in a new way as I listened to Jerome. Our road paralleled the pathway that the Old Salt Woman left as she walked away from Zuni, dropping her earrings as she journeyed. He pointed out a cave on the mesa where the Katchina lived that punished mischievous children. Places were knitted together with story, history, morals, and responsibilities. Everywhere was a reminder of who one was as a member of the Zuni community.

One bright spring day, a group of archaeologists had gathered to help me map a portion of a historic peach orchard on the side of the sacred Corn Mesa, Dowa Yalanne. Over the millennia, fine sand had settled on the leeward eastern side of the mesa as the harsh spring winds scoured the high desert plateau. In these sand dunes had been planted an orchard from rootstock brought to the Southwest by early Spanish priests. The deep sand, high elevation, and the southwest facing slopes provided an ideal ecological niche for the small shrubby peach trees. Now abandoned, the only indications of the peach orchard were stumps of the long-dead trees. I was mapping the locations of the stumps and field houses in the orchard to merge the data with a geographical information system to explore the relationship between aspect, slope, and planting patterns.

A Zuni high school student, Sheldon, was helping us that day with the transit and maps. He pointed to the red and white mesa above us and said, "You know, the water came up this high—those rings up there, that's from the flood." Prominent in Zuni oral history is a flood of massive proportions that required the inhabitants of the pueblo to retreat to the top of the mesa for refuge. For us archaeologists, we saw the story of the flood as folklore, but for Sheldon, the white rings were evidence and a mnemonic sign to a set of stories that outline the relationship between the Zuni people and their landscape. But more so, I understood how we non-Indigenous researchers spoke about time and history held different meaning for the members of the Zuni community.

American Agriculture and the "Hyper-Rational Hand"

Notes from my research are rich with examples of the everyday experience among farmers and gardeners in Zuni. Although the Zuni environment presented radically different challenges in cultivation from my own experience in the Midwest, the political and economic context at Zuni had also shaped my own farming community. I spent my youth in a small German-American farming community in central Illinois from 1959 to 1975 where my father was a large-animal veterinarian. Even though I was a "town kid," I divided my time between my best friend's home on a large farm and our "adopted" grandparents, an elderly couple who were part of a dairy farm just a mile from our house. I grew up with gardening, 4-H, and the village's rhythm of spring planting and fall harvesting, all dictated by a "common sense" that guided our activities.

Even as a young teenager in the early 1970s, I could tell that farming was changing in our community: my friend's parents took on extra jobs in addition to their farm; more families were converting from cattle to hog operations; and fewer families were able to afford veterinary care for their animals. The 20th century witnessed two periods in which American farmers lost their land through economic collapse and loan foreclosures: the 1920s during the Great Depression and again in the 1970s in what has come to be known as the Farm Crisis. In the 1920s American farmers changed the form and organization of their farms using components of industrialization; farms were re-conceptualized as a managed system that was modeled after factory production. Industrialization at this time was based on efficiency, specialization, speed of throughput, standardization, and a Fordist belief in rational management techniques. "Science, technology, and the spirit of rationalism that characterized industrial agriculture were created and maintained by a new class of people and institutions whose principal purpose was to modernize the whole agricultural enterprise" (Fitzgerald, 2003, pp. 5–6). In the 1970s, the Farm Crisis ramped up the industrialization to an even greater degree in order to compete in the growing global markets.

Coordinating the Transformation of American Agriculture

The key to this transformation of American agriculture in the 1920s was the coordination of policy and practice at national, state, and local levels. Federal, state, and county governments promoted "agents of industrialization" who were crop and farming specialists who consulted with farmers, organized clubs for girls and boys, held fairs and seed competitions, and taught classes in agronomy at high schools. At the same time, land grant universities with a new generation of agricultural scientists led research on improving agricultural seed stock and developing machinery for mass production. The industrialization of agriculture began with harvesting wheat, with an emphasis on recording yields per acre along with documenting the financial inputs of machinery, gas, fertilizers, seed, and other costs. This detailed quantification and documentation in standardized formats was essential to the rational management of agriculture. Behind the transformation of American agriculture was a set of ideas, practices, and interactions based on the "industrial logic for agriculture" (Fitzgerald, 2003, p. 8).

Fitzgerald noted that in the 1920s an agricultural leadership emerged: businessmen (bankers who funded loans for seed, livestock, and implements); county agricultural agents; and college (land grant) professors. These leaders developed an "industrial logic" and a "matrix of ideas" that persuaded farmers to change their practices, which resulted in transforming rural America. Among this new professional group were college-educated agricultural experts who became the authority on analyzing agricultural engineering and economics and whose intent was to modernize agricultural production and use a "hyper-rational hand" (2003, p. 27) in reorganizing the day-to-day tasks in farming life. This rationalization became the "common sense" that guided local practice and decision-making in farm households. For example, part of the push to modernize American agriculture was the Rural Electrification Administration, which brought electrical lines to farms across the US. "Now the barn not only could be electrified, it *should* be electrified, and the cows not only could be milked by machines, they *should* be milked by machines" (2003, p. 185).

As much as it was an agenda at the national level, the transformation of agriculture played out with different crops locally: Kansas wheat, Louisiana rice, California citrus, and Midwestern corn and soybeans. At the local level, the county cooperative extension service was critical to this transformation. Beyond the US, the Soviets' scientific approach to mass production in agriculture provided a challenge to the U.S. domination of the international grain markets. In turn, the rationalization of industrial agriculture and the analytical tools the Soviets and Americans promoted made agriculture transferable to other developing countries in what was known as the Green Revolution: scientific solutions and expert advice on entrenched agricultural problems.

4-H Clubs: Agents of Industrialization

Prominent in my youth was my participation in 4-H, a club in which girls and boys took on agriculture, gardening, livestock, and home economic projects over the course of a year. I remember taking 4-H very seriously when choosing my project, completing the logbooks, and recording "data" from my garden. The year culminated with 4-H club members showing our projects at the

Livingstone County Fair. I brought in carrots, tomatoes, and string beans from my garden one year and was thrilled that I won a blue ribbon, as well as a chance to show my garden produce at the State Fair in Springfield.

Not only were youth engaged in competitions at county fairs, farmers, too, brought livestock and produce to be judged by the county extension agents. My adopted grandparents brought their Guernsey dairy cows to the fair and were proud of the many ribbons they garnered (Figure 3.4). Winning ribbons not only was a source of personal pride, but also farmers who were breeding cattle or other livestock wanted to promote the quality of their herd, indicated by the ribbons. Of course, the agricultural extension agents and university agricultural engineers had developed a set of standards for the each breed, vegetable, and crop that were used in the judging process at the local fairs.

The 4-H movement was a response to early 20th-century migration of the rural population to the city and the subsequent urban industrialization (Wessel & Wessel, 1982). Albert B. Graham—a superintendent for Springfield Township,

Figure 3.4 The author (on the left) is holding a string of ribbons from the county and state fair for showing Guernsey dairy cows.

Ohio in 1902—argued that rural youth needed vocational training in agricultural and household management. In response, he organized the first agricultural clubs for the boys in which they used scientific techniques to test soil acidity on their farms and the germination rates of their seed stock. At the same time in Illinois, Will B. Otwell, president of the Farm Institute, organized corn-growing clubs for boys in 1901, along with a competition that awarded new farming technology to the winners: new ploughs and cultivators. Under the rubric of agricultural science for boys and domestic science for girls, these clubs distributed new seed and invited youth to display their produce at county fairs. The goals of the agricultural clubs revolved around improving corn production and seed selection practices by promoting "scientific agriculture."

Conversely, girls' clubs focused on canning, sewing, baking, and "domestic engineering." These girls' clubs did not share the technological goals of improving agricultural techniques. Instead, girls were positioned in the role of community responsibility. The clubs began as a place for girls to learn how to can tomatoes with materials sponsored by the U.S. Department of Agriculture to teach girls safe and efficient techniques for preserving food.

Young people in clubs were instructed to increase local food production and to demonstrate methods for substituting foods in cooking for ones deemed "less critical" thus helping to save supplies of red meat, wheat, and sugar during World War I and II. Early on, club members used a standardized reporting system for recording methods, data, and interpreting the results of production, seed trials, and plant growth. Standardization and reporting were essential to the agricultural clubs attended by the boys. After 1959, the Cooperative Extension Service worked with the National Science Foundation to deliver scientific content through 4-H. In the early 1960s, the clubs stepped up their scientific content and science process in the delivery of their programs.

Agricultural Change at the Zuni Pueblo

The hyper-rational hand of agriculture was also imposed on the agriculture at the Pueblo of Zuni. Agricultural clubs for boys and girls were never instituted, but the agricultural extension agent and more importantly, the Bureau of Indian Affairs (BIA) developed agricultural policy on the reservation that dominated daily activities. Despite the tribe's reputation in the Southwest of being expert farmers with several years of grain in reserve, the push to industrialize agriculture at Zuni became a federal agenda in the early 1900s as part of its "custodial" responsibilities of assimilating Indigenous communities into mainstream culture and economy. The federal government wanted to shift agricultural fields from the dry-land farming to the irrigated bottomlands on the Zuni River, where fields could be more effectively industrialized. In 1904, the BIA constructed a series of dams at Blackrock and other farming villages on the Zuni reservation. The Blackrock Dam failed within a year of its construction, scouring prime topsoil from arable land downstream.

Soon after the Blackrock Dam was reconstructed, a BIA school farm was initiated in 1912 to demonstrate mechanized agricultural techniques. The government meant to replace hand labor on the Zuni Reservation with mechanized agriculture to increase productivity. The outcome of this mechanized system was to radically change planting techniques. As a result of the new mechanization, crops were planted in large plowed fields rather than in small, scattered fields planted by hand with digging sticks. Plowing with tractors and draft animals exposed the fragile soils and made them vulnerable to erosion. Another consequence of mechanization and the dam projects was the break down of traditional systems of labor organization. Traditional work groups organized at farming villages were disrupted when the federal government encouraged individuals to accept 2 acre allotments in the Zuni Irrigation Unit.

Zuni farmers also switched from maize with a floury endosperm (one that is easy to grind between stones in a traditional way) to maize with a flinty endosperm that is used primarily for cattle feed. Even though farmers did not record production on their own, Zuni farmers were required to bring their produce to the BIA for weighing.

Despite their efforts to enter the regional markets, farmers at Zuni never had the economic capital to invest in machinery or the improved seed stock for industrialized agriculture. Nonetheless, the influence of the BIA and the agricultural extension agents had a profound affect upon Zuni agriculture as tractors and plows became standard. Also, many farmers converted their acreage to alfalfa and no longer grew traditional crops for local consumption. Alfalfa became the only marketable crop beyond the reservation borders. Still, in small pockets of the reservation, fields of traditional crops exist, largely due to the need for white and blue maize for traditional ceremonies.

A Consciousness of Everyday Practice

In her approach to understanding social and cultural behavior, Smith (1987) describes how the "ruling relations" are expressed through everyday practices. Our daily labor at work and home are connected to dynamic matrices of relations and production, regulated by the forces that emerge from it. This sociology is bent on developing a consciousness: discovering these relations, how they are ordered, the ways that they function at local historical settings of work and the ways that we encounter others. Smith defines ruling relations as a concept in which power, organization, direction, and regulation are interwoven, more than other discourses of power acknowledge. From a mode of ruling emerges a "transcription" (1987, p. 3) of locally specified activities that become abstracted and generalized as part of our larger consciousness.

In this chapter, I describe the hyper-rational hand of American agriculture and how it played out at my home in the Midwest and the Pueblo of Zuni, as an example of Smith's ruling relations. Scientific practice, measurement, and standardization were part of the forces of production that directed everyday practices among farmers and gardeners. As Porter (1995) has argued, a trust in numbers—as in the quantification of agricultural production—was an outgrowth of the social and political pressure to industrialize and standardize agricultural practices. Even our survey of traditional Zuni crop diversity included elements of the hyper-rational hand where we carefully carried out a census of participants with forms, catalogued samples, took photos, and produced our final report complete with charts, graphs, and basic statistics.

These years at Zuni have had a profound influence on how I now view traditional knowledge systems and Eurocentric science. In communities like Zuni, I now see the landscapes as being part of social memory, in which one's relationship can be mapped to larger systems of knowledge. Similarly, Davidson-Hunt and Berkes (2003) describe the ways that ecological knowledge is a co-constitutively composed, drawing from social memory (perception, cognitive knowledge, technology, institutions, and worldview) as participants move through their daily life (for an example of a similar "counterstory" see Richardson Bruna and Lewis, chapter 6, this volume). This knowledge is revisited through a communal act that requires a long-term understanding of place, environmental change, and cultural practices. Competency and authority in this social system is conferred through one's participation in the practices of daily life within the web of relationship rather than through the accumulation of specific knowledge. Also, I see how larger political and economic forces become normalized in daily practice and disrupt local knowledge systems. If we are to pursue a new "localism" in place-based education, we need Smith's approach to expand our consciousness of everyday practice.

Coda

Anselmo later came by to assess my sad garden at Zuni and immediately had a solution for me. We drove my truck down to a nearby arroyo where there had been a small flash flood earlier that week and he pointed out the thick piles of pine needles, cones, and fine sand. "Your garden needs this tree-sand," he said excitedly. We took out our shovels and filled the back of my pickup with the humus-rich sand, full of nutrients washed down the mesa. Anselmo also gave me seeds adapted to the Zuni summers instead of the plants selected for the ideal weather of the Midwest that I had been buying from seed catalogues. And finally, I rebuilt my garden into grid-like waffles, just as I had seen in historic photos of the old pueblo, where water would be directed on the roots of the plants, and also form a barrier to protect tender seedlings from the harsh wind.

In my succeeding years at Zuni, my gardens flourished as they have at every home I have had subsequently. Anselmo has long since passed away, his knowledge now practiced by his children and grandchildren who still grow his beans. Each spring I look for the first hummingbird before planting my beans, because I know now—the hummingbird never lies.

Notes

1 All names in this chapter are pseudonyms.
2 I refer to science as Eurocentric to indicate its European origin and to distinguish it from indigenous science. Aikenhead and Ogawa (2007) refer to science as Eurocentric sciences (plural) to emphasize the heterogeneity that exists among scientific practices and the multiplicity of paradigms in science derived from a European historic tradition, some of which are incommensurate. Yet, they also acknowledge that a set of common beliefs, norms, and values exist around Eurocentric science (singular).

References

Aikenhead, G. S., & Ogawa, M. (2007). Indigenous knowledge and science revisited. *Cultural Studies of Science Education, 2*, 539–620.

Bunzel, R. (1932). *Zuni rtual poetry. Forty-seventh annual report of the Bureau of American Ethnology, 1929–1932*. Washington, DC: Smithsonian Institution.

Davidson-Hunt, I., & Berkes, F. (2003). Learning as you journey: Anishinaabe perception of social-ecological environments and adaptive learning. *Conservation Ecology, 8*(1). Accessed May 18, 2008 at: http://www.ecologyandsociety.org/vol8/iss1/art5/

Fitzgerald, D. (2003). *Every farm a factory: The industrial ideal in American agriculture*. New Haven, CT: Yale University Press.

Porter, T. M. (1995). *Trust in numbers: The pursuit of objectivity in science and public life*. Princeton, NJ: Princeton University Press.

Smith, D. E. (1987). *The everyday world as problematic: A feminist sociology*. Toronto: University of Toronto Press.

Smith, D. E. (1996). Telling the truth after postmodernism. *Symbolic Interaction, 19*, 171–202.

Wessel, T., & Wessel, M. (1982). *4-H: An American idea 1900–1980*. Chevy Chase, MD: National 4-H Council.

4 Language and Experience of Self in Science and Transnational Migration

SungWon Hwang and Wolff-Michael Roth

In this chapter, we are fundamentally concerned with the mediation of personal learning experiences by transnational migration, one of the phenomena that comes with and constitutes the globalization of cultures. From rationalist perspectives, moving from one culture to another is not a problem for knowing and Self because it simply involves changing from one system of linguistic and cultural codes into another. All one has to learn is how to translate between the two respective forms of code. Within rationalist perspectives, such translation is taken to be unproblematic, as the proliferation and use of mechanical translators on the Internet shows. However, as we exemplify through our own experiences in this chapter, transnational migrations both within Western culture and from Eastern to Western cultures are associated with a substantial change of the bearings that allow a person to mark and remark sense; and it challenges the culturally specific forms of identity that we have been drawing on prior to migration (Roth & Harama, 2000). That is, transnational migration brings about a shift in everything that mediates who we are, that is, our identities. These identities constitute both a central resource in what we can and will do and the products of our actions, which *come* to be context- and experience-independent only through substantial work of abstraction. This also is the case for science and science learning.

In science learning environments, students are introduced to the new languages of the subject matter by means of everyday language, bound as it is to our cultural experiences. For transnational migrants, therefore, the teaching and learning of physics, for example, happens in a language other than that which has connected the person to culture and lifeworld forever, as far back as she can think. Learning science then becomes a matter of a double translation from the special language of the discipline (here physics) into the culturally dominant language spoken and used in the school or at the university (here English) and, for transnational migrants, from this dominant language and culture into the one that they are familiar with since birth (for the individuals in this chapter Korean, Japanese, German). This translation, as the Italian creators of the expression *traduttore, traditore* (translating is committing treason) have known for a long time, involves a substantial shift in reference points (e.g., semantic fields) and therefore inherently changes the ways in which sense is marked and remarked in everyday interactions.

In this chapter, we exemplify a way of understanding the shift in identity that occurs as an aspect of migration to different cultures and languages. In this endeavor we take a non-representational perspective of language, which means that language is *not* a medium for *re*-presentation but a way of knowing the world in and through each concrete communicative act, including speaking, listening, writing, and reading. Thus, knowing a language is the same as knowing one's way around the world. We follow Lev Vygotsky (1986) in the position that language is related to thought in an irreducible, whole unit of communicative performance; in this unit, language is only one constituent resource besides other sense-making resources. We look at learning science from the standpoint of transnational migrants, which is part of our own lived experiences—SungWon, Michael, and Miko have come to Canada more or less recently, and all three we have come to face the loss of familiar bearings as we confronted the foreign/strange in our new surroundings. Locating the chapter in our own experience of moving between nations and cultures, which involved shifting to speaking a language other than our mothers' tongues, we show that translation, and therefore the transition of identities, is the work that we, in and through our living (sensing, acting) body perform in communication. We suggest the *passivity* of the body—which we always understand as a *living* body that constitutes the source and place of the possibility of being affected, and therefore of learning—as the central concept for theorizing the lived experience of shift and the generalized possibilities that this shift involves for learning. As a way of articulating the issue associated with language and Self, we begin with an experience SungWon has had recently, which exemplifies how migration and language mediate the lived experiences in the new country and culture.

Language, Translation, and Self

We understand translation to occur not only between languages but also within one and the same language, which therefore is not one but many. For example, the different discourses and voices heard in the family context and those we participate in at work are marked by different societal-hierarchical ideologies (Bakhtin/Vološinov, 1973) and therefore means of producing and reproducing the understanding of others and self-understanding. This is so even if we were to deal with but one vernacular rather than with the different ways in which everyday culture and, for example, physics express themselves. Migration from one cultural and linguistic community to another brings to the foreground much more sharply than experiences within a culture and language (e.g., from one dialect to another) the instability of Self and the problematic nature of translation from one language to another. In this section, we make salient how transnational migration brings about opportunities and constraints for learning and for learning about Self.

> SungWon: I am on my way back home from Chicago. The plane is flying into Toronto, where I am scheduled to transfer to another flight bound

for Victoria. I am sitting in the center of a three-seat row. A woman, who sits left of me, talks to a flight attendant passing by and asks whether she has to claim baggage in Toronto. Because I have the same question, her words immediately catch my attention. The flight attendant tells her that the captain will make an announcement later on. Hearing the conversation, I find that she, too, is scheduled to transfer to another flight in Toronto. I ask her where in Canada she is heading to and tell her that I am going to Victoria. She says that she has been to Victoria for a whale-watching cruise and then continues the conversation.

The plane prepares to land in Toronto. I hear the captain announce: "passengers who transfer to flights outside of Canada . . . passengers who transfer flights within Canada . . ." I attend to the announcement and soon find that I missed the point. I am not sure whether I have to claim my bag or not. I turn to the woman on my left and ask, "Do we have to claim baggage?" Instead of talking to me, she looks around and talks aloud to people about what the captain has said. I hear someone repeat the captain's announcement and now I know that I need to claim my bag.

Soon, the plane lands in Toronto. After getting off the plane and on my way to claim the bag, I do not feel comfortable about the conversation that I have had with the woman. I feel that my talk might have sounded strange to her because of the word "we." I do not remember exactly what I said at that time, but I am sure I began my question by saying "do we" rather than "do I." I do not feel comfortable about the word that I said ("we"). I think that I was not "traveling with" her. I merely sat next to her and, at one point, had a brief conversation. I think that my questioning might have made *her* feel bad. I said "we" because I knew that she too was wondering about having to claim her baggage prior to going through customs and immigration. I have a sense that saying "Do I . . .?" or "Do you . . .?" is inappropriate because it sounded like separating what belonged to me and what belonged to the other and therefore produced a split in our *common* condition that I was aware of in that situation. I find that this is not the first time that I felt inappropriate about my using "we," "us," or "our." I think that it is because of differences between the English and the Korean language, particularly in the use of the word "we," which I used to say to others when I experienced this issue.

There are differences between Korean and English. However, before I reduce the issue to cultural differences, I recognize that I can find similar cases even when I speak in Korean: translations occur even within a language, which therefore cannot ever be identical to itself.

If the problem is not limited to the matter of language differences, it is worthwhile to think more about my feeling inappropriate. Why did I have

> to feel uncomfortable? Why did I have to feel otherness with respect to the woman next to me despite the shared experience of traveling on the same plane, sitting side by side, and having a conversation about trips? Why could I not have said to myself, "Why not 'we'?" and stopped feeling bad? I find that at the center of my feeling lies the shifted Self that I now experienced as (culturally) illiterate in contrast to the literacy of others. My feeling uncomfortable shows that I am confronted with my Self, which was overly attuned to the lack of cultural literacy and therefore to the literacy of my neighbor. Perhaps this is why I was attuned to "we" more than to any other word so that I do not even remember now why I felt uncomfortable when I saw her asking others, not answering to me directly. Rather than just listening to her without having to be aware of any otherness, I was thinking about myself as having missed the announcement and the foreignness of the English language. By attributing my discomfort to my lack of fluency in and with the English language, on the one hand, and language differences on the other, I did not give myself room to take another look at the situation, and thereby reflect on the way I experience my shifted Self.

Migration and the exposure to different cultures and languages mediate and (literally) *affect* (from Lat. *afficĕre*, to do, act on, influence, attack with a disease; f. *ad*, to + *facĕre*, to make, do) the Self. We know from our own experiences in science lectures that even the best of students may experience forms of estrangements that SungWon here articulates in terms of experiences on the plane: we sit in science classes, listen, and feel completely out of place. Emotions that work in such moments (e.g., satisfaction, disappointment) are expressions and constituents of the Self that experiences itself in transition and change. The transition and associated emotions are problems of the Self if we understand identity to pertain to re-presentations of the *same*. If, on the other hand, we understand identity in terms of non-self-identity—that is, as a construction that reduces what is evidently different to making it the same—then the experience of a shift denotes the inherent heterogeneity and plurality of the Self.

The vignette exemplifies shifts that being (living) in a different culture and speaking a different language have brought about for SungWon. First, she has had a conversation with her neighbor on the plane. She actually could have been happy about having a neighbor who had the same concerns and therefore asking for help, but she experienced it as something inappropriate. There was a difference between the Self that she concretely articulated in the act of talking to her neighbor on the plane and the Self that she perceived while reflecting on her utterances. The difference comprised two moments, the event itself and its subsequent image on which SungWon reflected. SungWon's sense of inappropriateness was another form of experiencing the Self, ex-scribed in and by means of her words, with which she had addressed the other woman. In this respect, the Self realized in feeling inappropriate has a discontinuous aspect from that which, in talking, she ex-scribed and articulated—the temporal difference constitutes an aspect of this discontinuity as the sense of inappropriateness emerged after she had finished asking her question.

Second, SungWon's narrative exemplifies a case in which the non-self-identical aspect of the Self (the one that we denote by typing "SungWon" or by uttering sounds that others hear as "SungWon") in the two events is reduced to the disapproval of the Self (illiteracy) or the otherness of the Other (strangeness of English language and culture). By attributing her sense of inappropriateness to language differences, SungWon produced and reproduced herself as an illiterate person. In feeling inappropriate, SungWon lost the flow that characterizes normal, everyday, and unproblematic way of *being-in-the-world*; instead, she entered an experience of otherness and estrangement. This feeling of being there but being disconnected from the world, which appears to play a secret game to which the migrant is not privy, is precisely what the individuals in this chapter have experienced following migration to our new country. And it is also an experience that characterizes students in a (physics) course, where the professor is denoting entities and things in or from a different world.

The feelings that SungWon experienced are not what SungWon (could have) intended: they impinged on her from elsewhere, af-fected (did something to) her, and left her vulnerable. Of course we can say that this experience was the result of migration and speaking another language (English), one that is other than her mother's tongue (Korean). But such a framing does not get to the heart of understanding the experience from a standpoint within it. In her narrative, SungWon articulates a sense of inappropriateness, and she explains the experience to herself in terms of language differences. She thereby sets up language as a pure object. For SungWon, both English and Korean gained purity. When she realized that the sense of "we" does not lie in the languages themselves, she came to an understanding (in a narrative form) of the shifted nature of her Self and her feeling. In this, SungWon was no longer Korean and was not yet Canadian. (We are aware of the fact that this framing itself is problematic, leaving open the questions of whether anyone can *be* something like Korean or Canadian and leaving open what it means to be *Korean* or *Canadian*.)

Transnational migration, more clearly than other cultural experiences, shows how every event changes us forever and shows how we always depart and never arrive. It shows how our identities remain unfinished projects, always in a transition between a past-having-been and a future-not-yet. And this is so because the Self not only is active in transnational migration (deciding to move, moving) but also, and essentially so, is *passive* (one's body has to be moved, the new and foreign *impinges* on the body). SungWon's narrative shows that the Self is *passive* with respect to upcoming possibilities that it realizes in action. SungWon says what she says (e.g., "we"), that is, a word is spoken in the fullness of the situation (the Self) of which the reduction to the English or the Korean always is a limiting abstraction and does not fully explain *why* she says what she says. Once the action is performed, the Self is *passive* with respect to the currently realized and yet already past possibilities. They constitute the conditions for the Self to be (radically) confronted with itself—as an Other. SungWon's articulation of her sense of the shifted Self arose from being open to be acted upon (i.e., radical passivity), rather than from an abstraction (e.g., language difference) that has been reduced from, and therefore is partial to, the lived experience as a whole. Identities change in translation—we learn to speak and learn about the Self—in the concrete performance (speaking, writing) in which we are affected, mark sense, and remark sense.

In what follows, as a contrast to the first-person perspective in the previous section, we take both first- and third-person perspectives to look at a case of science learning following transnational migration. We extend our non-representational and non-reductionist approach to language and the Self; we articulate to science aspects of passivity that are made salient following migration, and exemplify the central role of the living (sensing) body in the cultural translation; and we take an example of learning physics that Miko, a student who had come from Japan to Canada to study geography and meteorology, recently experienced as part of her undergraduate study, and analyze conversations about physics problems that Miko has had with a classmate.

Double Translation in Learning Science: The Role of the Body

The phenomenon of speaking a language that a person has not spoken so far and therefore cannot even imagine what it would be like to speak—e.g., learning to speak the canonical discourse of physics—highlights an aporia that learners are essentially *passive* in relation to what they come to learn and what they come to be (Roth, 2007). We cannot *intend* learning a specific thing (fact, concept, theory) unless we already know this thing: intention always has an object, and this object has to be given to the conscious subject. But if we already knew the thing that we were supposed to intend to learn, we would no longer need to learn it. This is the essential learning paradox framed in the form of a contradiction: we are asked to learn something that we cannot *intend* to learn. The exit from this paradox lies in this: both the intention and the learned object give themselves to us, individuals who play the welcoming hosts of (or are hostage to) what we had not intended.

This radical passivity is a central and constitutive aspect of learning, not because we are not knowledgeable (literate) but because what we come to know is irreducible to what is already known to us. For example, the specifically scientific discourse that we learn while attending a science course cannot be explained on the basis of the non-scientific discourse that we bring to the course. Learning science, as the term scientific *literacy* implies, means learning to speak a new language together with practicing what members in the field normally do. For transnational migrants, this is even more salient. In schools and universities, the foreign concepts of science are articulated and explicated in the culturally dominant vernacular, the language of the host country; but as migrants, we have to translate this vernacular into the language that we are most familiar with, generally our mother's tongue (Roth, 2005)—though for immigrants to some cities and countries, like Montreal in Canada, this may often be a third or fourth language. As migrants, we are affected not only by the different ways of knowing the world in the new country (see Richardson Bruna & Lewis, chapter 6, this volume) but also by the strangeness of the language of science that we learn after arriving there. This experience of being affected occurs throughout our normal everyday school lives in which we are intensively engaged with cultural practices such as listening to physics lectures and taking notes, conducting experiments in the laboratory and writing reports, solving physics homework problems, and completing answer sheets during examinations.

Any native speaker can reproduce the experience of strangeness by reading a research article or specialized book on a topic outside one's field: we read the words but know that we do not get the point. It is as if we walked on unfamiliar terrain, completely disoriented and without bearings. Something essential seems to bypass us, even if we know every single word; we may attribute our disorientation to the "jargon" used by the author(s). We may experience the language of the others as impenetrable, even when the texts only employ familiar terms in our mother tongue. That is, learning—the one that follows transnational migration specifically and all learning more generally—constitutes experiences that intermingle the familiar and the strange, disorient us in a world that nevertheless appears to be accessible and sensible.

In this section, we take a look at how, in those everyday works of translation and learning science, the identities of transnational migrants are in transition. We take examples from Miko's university physics courses, in which she experienced boundaries between herself, on the one hand, and physics and the English language, on the other; she did so in ways not unlike the way SungWon felt *being* othered (passivity) while *actively* using the English language that nevertheless is so foreign to her. We show that passivity and its source and ground, the body, perform translations of languages and make possible the transitions between identities. We exemplify the transition of identities in learning science with two cases of translation: first, the translation of disciplinary languages of science into languages that articulate students' scientific knowledgeability and, second, the translation of institutional languages into language that articulates students' lived experience of learning science. The two cases constitute an example of double translation of (foreign/strange) languages that transnational migrants encounter while learning science in a foreign land.

The Disciplinary Languages of Science are Translated in Passivity

Various linguistic resources mediate what and how we learn science. For example, undergraduate physics students solve physics problems with reference to textbooks, notes, and other available sources. All those materials are resources that provide us with exemplary ways of speaking and writing physics. In the act of reading and interpreting those resources for the purpose of solving the problems at hand, we, in and through our bodies, are affected by disciplinary languages of science and engage in translating them into a concrete realization of scientific knowing—any articulation (speaking, writing) is a translation that arises from the irreducible relation between the reader and the read, in which the passivity of the body takes a central role.

As part of undergraduate studies, Miko and her classmate Cathy frequently organized informal problem-solving sessions designed to assist one another in their attempts to solve homework problems in physics. As usual, the two have brought to the session whatever they had done individually. In the following episode, Miko and Cathy talk about a page of an engineering textbook that Cathy suggests as a useful reference for solving a thermodynamics problem (see Figure 4.1a). They work to prove physical properties of the Otto cycle—network, volume, pressure, and efficiency—assigned with a diagram in the problem sheet (see Figure 4.1b).

Episode 1[1]

01 C: oh, no,
[$_1$r: was the
[$_1$((Cathy points at a text in the book. Miko moves the problem sheet closer to the textbook while gazing at the textbook.))
02 [$_2$um:
[$_2$((Miko turns her gaze from the textbook to the problem sheet.))
03 [$_3$(1.2)
[$_3$((Cathy flips her right hand over and makes the palm curved.))
04 [$_4$compression ratio that's it
[$_4$((Cathy points at the textbook and Miko moves her gaze from the problem sheet to the text that Cathy is pointing at.))
05 M: [$_5$o:kay
[$_5$((Miko points at the textbook with her left hand.))
06 ((Miko brings her problem sheet closer to the textbook that Cathy is pointing at. Cathy moves her right hand away.))
07 [$_6$so in this ca:se
[this two, too. ((Miko points at the diagram in the problem sheet with a pencil on her right hand. She moves around the pencil above the diagram.))

Cathy says "oh, no" and, with her right hand, points at equations on the page (line 01, Figure 4.2a). She produces an "r" that is longer than usual. This "r" is a physics notation that the two have talked about to figure out what it stands for. Cathy continues, "was the" (line 01). She thereby articulates a conversational

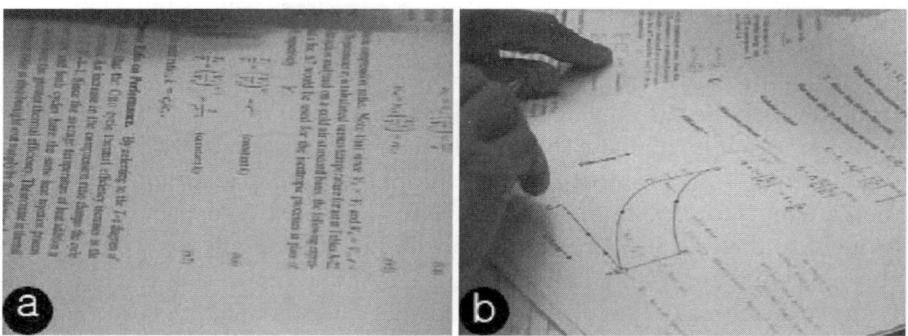

Figure 4.1 (a) Cathy's engineering textbook shows equations and (b) Miko's problem sheet has scribbled notations and equations around a diagram.

topic. Miko moves her problem sheet closer to the textbook (line 01, Figure 4.2a). Miko gazes at the textbook (line 01). Cathy utters "um"—an interjection that is used to indicate hesitation on the part of a speaker and Miko turns her gaze from the textbook to her problem sheet (line 02, Figure 4.2b). A 1.2-second pause follows. Cathy flips over her right hand and bends the palm to make it curved (line 03, Figure 4.2b). Cathy points to the textbook (Figure 4.2c) and articulates "compression ratio, that's it" (line 04). Miko turns her gaze back to the textbook (line 04, Figure 4.2c). Miko says "okay," which others might hear as an agreement (line 05). She points at the part of the textbook that Cathy is pointing at (line 05, Figure 4.2d) and then moves her problem sheet even closer to the textbook while leaving her left hand pointing to the textbook (line 06, Figure 4.2e). Cathy takes her right hand away from the textbook, which thereby leaves Miko's hand pointing the equation by itself (line 06). Miko continues to say "so in this case" (line 07). She moves a pencil in her right hand to the diagram and point to a part of it (line 07, Figure 4.2f).

In this situation, we see Miko and Cathy engaged in interpreting equations one page of the engineering textbook (Figure 4.1a). The conversational topic is "r," the notation that they have found in the textbook equations. Cathy orients her body to the textbook page and utters "r" simultaneously with a pointing gesture to the equations. The body orientation, the speech, the pointing gesture, and the material text ("r") constitute a communicative unit that makes the issue available to both Miko and herself, and thereby constitute a translation of the concept that the text "r" refers to. Cathy continues ("was the") and hesitates ("um"). Her utterance does not move on. Instead, she redirects her pointing gesture to a curved shape and holds the hand over the textbook page. This gesture and the text beneath it together thereby constitute another form of translation—this is a momentary concentrated unit of a *growth point*, from which thinking and the utterance, "compression ratio," and a pointing gesture emerge and develop. Cathy's series of actions exemplifies that articulating scientific knowing is not the literal change of one word to another but

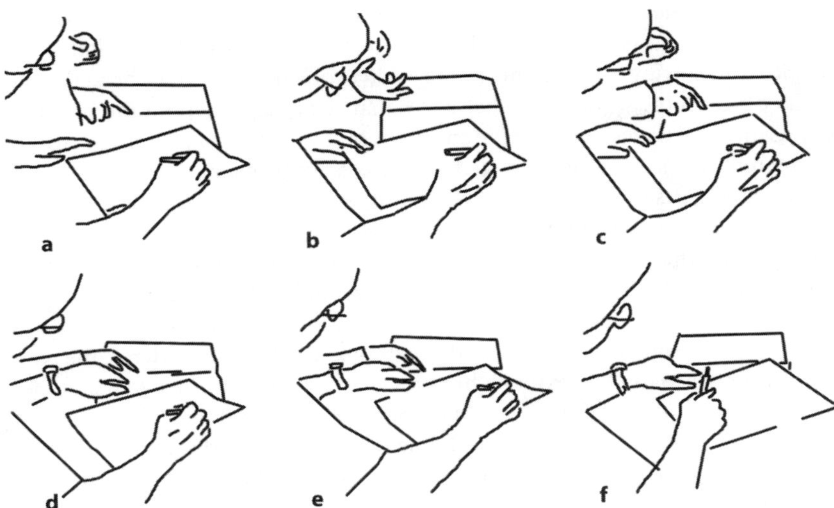

Figure 4.2 (a) Cathy points at an equation in a textbook and Miko gazes at it; (b) Miko turns her gaze to her problem sheet and Cathy flips her hand over; (c) Cathy points at the equation again and Miko gazes at it; (d) Miko points at the equation besides Cathy's hand; (e) Miko moves her problem sheet closer to the equation; (f) Cathy moves her hand away and Miko points at the diagram with her pencil.

constituted by continuous translations of communicative forms such as speech, body orientation, eye gaze, and gestures.

Throughout Cathy's turn, Miko's attention shifts between the textbook and the problem sheet. Miko gazes at the textbook and moves the problem sheet closer to it. Cathy stops speaking and Miko turns her gaze to the problem sheet. Cathy utters "compression ratio" and Miko turns back to the textbook. Here, Miko's changing body movements (eye gaze, body orientation) exhibit that her orientation toward the textbook is grounded in and occasioned by the physics problem on the sheet. This is articulated even more clearly in the subsequent movements. Miko takes a turn by uttering "okay" and she points toward the textbook. The position of her left hand on the textbook (pointing gesture) and her right hand on the problem sheet materially articulate that she is engaged in translating the two materials. The action of moving the sheet closer to the textbook, the speech ("so in this case"), and the change of the right hand into the pointing gesture with a pencil all exhibit the body in the process of performing translation.

In this episode, we find two forms of passivity are central to the ongoing translation. First, Miko's series of body movements shows that she is open to, and thereby being affected by, a new approach that Cathy suggested. This is one of the under-theorized aspects of learning: we encounter the unfamiliar, foreign, and strange, which, because inherently we cannot anticipate it, we have to allow it affect us. At the beginning of this problem-solving session, Cathy said to Miko that she found the equations and diagrams in her engineering textbook that are helpful for solving one of the homework problems—she has brought the textbook and now shows Miko her notes that she had written on the previous day. Cathy articulates for Miko the approach in the engineering textbook, which is different from the steps suggested on the physics problem sheet and she still has missing parts in her calculations. That is, it is not completely clear how the textbook helps in doing those calculations that they have to produce to prove the equations on the problem sheet (Figure 4.1b). Miko and Cathy both stare at the textbook. They do not even discuss whether it would be a good idea to follow the approach in the engineering textbook. Rather, the two attempt to figure out solutions following the approach in the textbook. For the two students, the issue is not whether to use the textbook but how it is related to the problem that they have to solve. Miko already accepts the work that Cathy is doing (interpretation of "r") while she gazes at the textbook: she moves herself and the sheet closer to the textbook, iterates Cathy's pointing gesture (double pointing to the textbook), and substitutes it in consequence (two simultaneous pointing gestures to the textbook and the problem sheet).

Second, passivity involved in the openness to a different (engineering) approach allows a critical reading of both the textbook and the physics problem, and thereby constitutes a ground on which a different form of talking physics is articulated. Throughout the session, the discussion of the two is focused on how different ways of talking are related to one another (e.g., different ways of talking Otto cycles) rather than privileging one form of talk. The discussion concretely realizes some of the generalized possibilities that different forms of talk provide about the understanding of the phenomenon (e.g., Otto cycle), which ultimately turn out to constitute a solution that is different from the exemplar that the professor has given them on the problem sheet. In not considering a different approach, the two students are passive with respect to the articulation that they produce in the continuous work of translation. The exchanges between the two students thereby exemplify that the (radical) passivity of the body constitutes an aspect central to translating disciplinary languages of science into those that articulate the scientific knowledgeability of learners.

The Institutional Languages of School Science are Translated in and Through the Passivity of the Body

It turns out that the problem-solving session contributed to improving the two students' understanding of physics concepts, as we exemplify in the previous section, and thereby increases their room to maneuver in physics—for example, Miko would be able to submit her homework when it is due and find time to go to the academic seminars that she really is interested in. This lived experience of doing physics provides opportunities to articulate their ways of talking about learning experiences. In the session, which usually occurs in the lounge of physics building, the two students talk about their university lives, such as what they feel about homework, classes, and courses. Those talks have recourse to institutionally dominant discursive resources, and thereby constitute aspects of the Self that are located in the societal order of institutions (Smith, 2005). That is, our ways of talking about experiences are affected by our institutional lives (e.g., undergraduate university schooling) and at the same time, our Selves are produced and reproduced in those forms of talks. Therefore, when the specific situation of learning physics involves a change from the dominant institutional forms of talking to those forms that articulate the lived experience of learning, the change would exemplify an aspect of our Selves that is affected by, and therefore passive with respect to, our lived experience. Our video analysis reveals that the passivity of the body is central to this articulation, performed on the material ground of what we have been doing, and thereby translates the institutional discourse of schooling and school science to those that articulate the lived experience of learning science.

Episode 2

11 C: [$_1$just under 50% for the coursework and just over 50%
[$_1$*((Miko presses keys on the calculator while moving her gaze between her answer sheet and the calculator.))*
[$_2$for them
[$_2$ *((Miko turns her face to Cathy.))*

12 M: [$_3$um
[$_3$ *((Miko lightly nods.))*

13 C: [$_4$like fifty three and a half percent for the final
[$_4$ *((Miko continues to nod.))*

14 M: [$_5$um
[$_5$ *((Miko nods more strongly.))*

15 C: [$_6$if he only gives us (assigned?) assignments
[$_6$ *((Miko continues to nod.))*

16 [$_7$ (2.7)
[$_7$ *((Miko keeps nodding and changes her gaze to the calculator.))*

17 M: [$_8$I like doing assignment
[$_8$ *((Miko presses keys in the calculator. Her speech marks a higher pitch value than the normal. Cathy moves her left hand to her notes.))*

18 C: [$_9$I do too
[$_9$ *((Miko presses keys in the calculator. Cathy's speech marks a higher pitch value than the normal.))*

19 M: [$_{10}$yeah:
[$_{10}$ *((Miko nods while continuing to use the calculator.))*

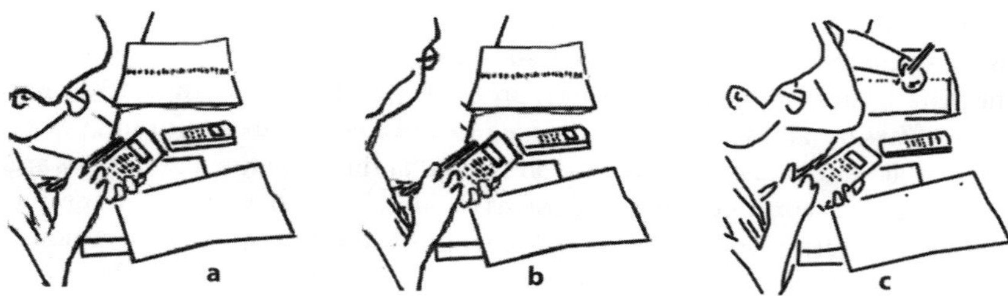

Figure 4.3 Miko (front) and Cathy (rear) have a conversation: (a) Miko is looking at her calculator; (b) Miko turns toward Cathy; (c) Miko turns her gaze back to the calculator.

Miko looks at the calculator that she holds in her left hand and presses keys with her right hand (Figure 4.3a). Cathy suggests that the final grade of their physics course would be counted by "under fifty percent for the coursework" and "over 50% for them [the final test]" (turn 11). Miko turns her face to Cathy (turn 11, Figure 4.3b) and utters "um" with light nods of her head (turn 12). Cathy continues to articulate "like fifty three and a half percent for the final [exam]," and thereby provides an example of her anticipating ratio (turn 13). In the meantime, Miko continues to nod (turn 13) and utters "um" again, associated with more extensive nods (turn 14). Her nodding continues as Cathy articulates the condition of her estimation, that is, he [the professor] gives the initially assigned number of homeworks (turn 15). All the while, Miko nods (turn 15). For a brief moment neither of them speaks (turn 16). Miko stops nodding and moves her gaze to the calculator (turn 16). She begins to press keys on the calculator and Cathy puts her arms on the table (Figure 4.3c). Miko says that she likes doing assignments (turn 17). Cathy immediately utters that she likes it too, and thereby articulates her agreement (turn 18). Miko utters "yeah" while nodding and looking at the calculator (turn 19).

In this situation, Miko and Cathy have an opportunity to talk about the physics course for which they are doing homework. They articulate for one another how much of their final grade will be based on the coursework, which includes their homework and the final exam. Cathy presents her estimate and Miko participates in the talk by the act of nodding and producing interjections ("um"). The conversation thereby provides an opportunity to articulate the value of their problem-solving work in terms of the ratio that it takes in the final "grading," that is, in terms of the dominant institutional discourse that characterizes university schooling. However, after a 3-second pause the two engage in a way of talking quite different from what they have done previously. The two suggest to each other that they like doing assignments (turns 17–18), and thereby provide a description that takes a standpoint of their lived experience of doing physics, not reducing it to the institutionally given criteria (grade). Therefore, their utterances constitute translations of institutional forms of talking into that which acknowledges the value of their work from the perspective of the experiencing person—here, institutional processes (homework, grading) constitute resources for learning and understanding the Self.

This conversation has come about not upon someone else's request but as part of their embodied problem-solving work. Miko's body engages with the calculator and the answer sheet on the one hand, which constitute a local space where the physics homework and the calculation for it are performed. On the other hand, the body is oriented toward the other speaker and accomplishes communicative productions (speech, gaze), which constitutes another space where the narrative (institutional discourse) is organized around the topic of grading.

Our description shows that Miko's utterance—that she likes doing assignments—is produced precisely when she moves her gaze from Cathy to the calculator and continues her calculation work with her answer sheets (turn 17). By orienting her body toward the calculator and notes, Miko opens and exposes herself to the physical space in which her embodied problem-solving work with Cathy is present. The body thereby is exposed to, and affected by, the understanding that those physical arrangements (in which the body is a constitutive moment) involve, because the world is comprehensible because the body has the capacity to be present to what is outside itself (Bourdieu, 2000). Miko's utterance marks a transition from the institutional discourse into the mode of (embodied) understanding of her work and the discourse that articulates it. The increase of the pitch value before and after the silence constitutes an indication that her utterance is not in a synchronic relation to her previous utterance and takes a different standpoint from it. Cathy has already changed her position and is moving her body closer to her notes when she produces an utterance that is in agreement with what Miko has said—the pitch value of this utterance, too, is significantly higher than her previous talk and constitutes an aspect of transition (translation). Thus, the situation expresses that the two are being affected together, which is a concrete realization of possibilities that their embodied problem-solving work has made available to them. The passivity of the living body—i.e., the body which not only acts but also is affected—allows the two to be exposed to the practice (i.e., the understanding distributed over the things that the body engages with), and thereby perform the articulation in the act of changing eye gaze, body orientation, and prosody. Therefore, this conversation exemplifies the central role of the body in articulating the lived experience of learning science, that is, the work that the two students are doing in the here-and-now of their situation.

This form of passivity, together with the passivity that allows the articulation of the two students' knowledgeability (as exemplified in the previous section), provides opportunities to critically engage with institutional discourses and translate them on the empirical (material) ground (i.e. critical literacy)—we are literate insofar as we engage with different ways of talking as different resources for articulating our lived experiences and therefore our learning identities. Miko and Cathy continue the conversation about the benefit of doing homework to improving their understanding of physics and augmenting their individual experiences. The two are (critically) literate in this talk because of their capacities to integrate their everyday languages with their ongoing physics problem-solving work and to hybridize culturally dominant languages (e.g., institutional discourse at university) into those that fit with embodied understanding. This critical translation of languages consists of a simultaneous double movement between the two kinds of sense that the passivity of the body allows: sense(s) that is given to the body in being-affected and the birth of sense from being-affected. Thus, when Miko and Cathy turn their bodies to the space of their embodied understanding and articulate that they like doing assignments together, there is a double movement of translation that marks and remarks sense: (a) a movement from doing homework as institutional work (sense) to the specific articulation of doing this homework that I am doing (senses) on the one hand and (b) a movement from the homework that they have been materially doing (senses) to perception of the work as a specific form of doing homework (sense) on the other. The two concurrent processes drive the (critical) translation of languages and the transition of identities, which is non-self-identical to itself.

Conclusions

The purpose of this chapter is to understand the experience of transnational migration and learning science from a standpoint of the person in transition, which, rather than being completed when she arrives in the new country, is an ongoing process in the course of which the individual hybridizes two cultures in and through her body. We provide our own experiences of learning to speak (foreign/strange) languages and exemplify the role of the body in the hybridization (translation) of languages. We take a non-representational and non-reductionist approach to languages and identities, and articulate the passivity of the body, which constitutes a central aspect of the translation of languages and therefore the transition of identities. This passivity offers opportunities to understanding, but it also brings about the estrangement we experience in new countries (here Canada) and new communities (here physics). To sum up, there is a double experience of passivity: on the one hand, my body makes available and is exposed to heterogeneous action possibilities, which derive from different (sub-)cultures and languages and which are affected by the latter. I perform actions differently than prior to migrating between nations, and this transnational migration is a good analogy for engaging in learning school subject matter that is foreign to me in my main language. On the other hand, my body is being exposed to the (being-affected) action that it materially performs, and thereby is affected by new possibilities that spring forth from previous concrete realizations of possibilities. A different sense of my Self arises from it. The double passivity that we experience and exemplify in this chapter explains why and how we learn and change not only in transnational migration but also in learning science.

Given these findings, we are aware again of the problem of approaches that presuppose languages as objects independent of the situation that I am subject to and my embodied experiences. Our examples in this chapter significantly show how languages can be mediating resources for communication, for learning, and for our learning identities only in and through the work that our living bodies perform. I am the subject of speaking (writing) and therefore learning because my living body is radically affected by languages that I agentially engage with and is host and hostage to their translations—this thereby leads to my Self that marks and remarks sense differently from the way in which it has been done before (double passivity).

Following phenomenological sociology, we articulate the openness of the living body to the world that includes the body itself and therefore allows us to be outside of our Selves all the while we feel them from within. The living body that is exposed to itself brings about a double passivity and estrangement, which therefore mediates the transition of identities that are not reduced to either position (e.g., *field*) or disposition (e.g., *habitus*) but are constituted by non-deterministic dynamics of the two sides. In this chapter we exemplify this hybridizing transition in the way it is grounded in the ongoing work (e.g., problem-solving session, writing the chapter) on which our embodied understanding is materially built up, not an abstracted transcendental movement.

Acknowledgments

This work was supported by a grant from the Social Sciences and Humanities Research Council of Canada.

Note

1 In this transcript, the following transcription conventions are used:

[<u>compression</u>	Beginning of overlapping talk or gesture, which extends for as long as the words are underlined;
((*Cathy points*))	Italicized words within double parentheses constitute transcriber's comments on visible body movements;
um:	Lengthening of a phoneme is indicated by colon;
(2.7)	Elapsed time in tenth of a second.

References

Bakhtin, M. M./Vološinov, V. N. (1973). *Marxism and the philosophy of language.* Cambridge, MA: Harvard University Press.

Bourdieu, P. (2000). *Pascalian meditations* (R. Nice Trans.). Cambridge: Polity Press.

Roth, W.-M. (2005). *Talking science: Language and learning in science classroom.* Lanham, MD: Rowman & Littlefield.

Roth, W.-M. (2007). Theorizing passivity. *Cultural Studies of Science Education, 2,* 1–8

Roth, W.-M., & Harama, H. (2000). (Standard) English as second language: Tribulations of self. *Journal of Curriculum Studies, 32,* 757–775.

Smith, D. E. (2005). *Institutional ethnography: A sociology for people.* Lanham, MD: AltaMira Press.

Vygotsky, L. S. (1986). *Thought and language.* Cambridge, MA: MIT Press.

5 Reality Pedagogy

Hip Hop Culture and the Urban Science Classroom

Christopher Emdin

I spent the most memorable moments of my youth listening to hip hop music through a portable tape player with oversized stereo headphones. I would wrap the long cord of the headphones around the tape player to keep its broken cassette player door in place and would hold the door shut with my hands to keep the music pouring through the headphones. Over time, the constant use of the tape player caused the headphones to develop an electrical short, leaving me with music only coming through one earphone port and forcing me to constantly twist the cord until I could bend it at the perfect angle that would allow the music to flow through both ears. The more quickly I could get the music drifting through both headphones, the shorter the time it would take for me to get lost in the sounds and words that poured through the player, and the easier it became to find solace in a world, where being from a certain neighborhood or of a certain race and class meant being accepted and loved in some worlds but almost invisible in others.

I have been in situations where police unlawfully searched my friends and me or when I felt ostracized by a teacher while trying to engage in a classroom discussion. In such situations there was always a rap song that either explained the situation or gave me comfort in knowing that the feelings that came from my experiences were shared with many before me and are currently being experienced by many more just like me.

As I grew older, my experiences growing up in an urban area became more challenging. I was introduced to worlds outside of my immediate one and the looks of contempt and feelings of being unaccepted that I got when I walked through certain areas introduced feelings of not belonging in the world beyond my neighborhood. When I entered secondary school, negative experiences that I had with some teachers and school administrators caused me to feel separate from even the science classes where the subject matter captured my interest. In response, I returned to my tape player and headphones and looked to them to provide an escape from my negative experiences. Daily, I would listen to hip hop music and find solace somewhere between the beats and the rhymes until eventually that was not enough. I continued to feel like I did not belong in school and that I needed to become more fully immersed in something that accepted me. In response, I moved from needing to listen to hip hop to wanting to create the same type of music that inspired me. I needed to become a part of the culture that inspired me and that had become important to my survival.

In the process of becoming a full participant in hip hop culture, I would meet with many of my peers who shared the same passion for the culture as I did. We all shared similar experiences, attended similar schools, and dealt with the same issues inside and outside of our neighborhoods. We would congregate at any place we could, gather in a circle and rap to beats produced by rhythms created with cuffed fists and movements of our lips. The words and sounds we created would take us away from whatever our immediate struggles were and would connect us to each other.

Becoming a Part of Hip Hop Culture

For students of color in urban areas, hip hop serves as an escape from the struggles of their everyday lives. Lyrics to rap songs tell a tale of both the physical realities of life in the inner city and the emotional frustration that comes with being ostracized from and silenced by mainstream culture. There is a deep relationship between rap and the lives of those oppressed in the inner city that those involved in hip hop deeply understand. In other words, the streets speak to the music and the music reports back what it hears from the streets. In some respect, all variations of hip hop fall within this framework. The music that is created serves as (a) a critique of the fields where we navigate descriptions of both the negative and positive aspects of our lives within these fields, or as (b) fantasies of what breaks in our temporary existences would be like.

For those who are not actively involved in hip hop, there is a belief that the hip hop generation is the embodiment of all the negative aspects of urban life and the underbelly of the nation's culture. From my experience I know that they fail to realize that hip hop is a product of a lack of voice in schools and various political and social arenas. In addition, there is little understanding that the pain, anger, and frustration expressed in hip hop are the collateral results of the feelings generated from experiences in the hip hop students' lifeworlds, rather than unfounded exhibition of negative emotions. Finally, it should be recognized that it takes spectacular ability to transfer experience and emotion into insightful words replete with analogy, alliteration, and vivid description—some of the many attributes that urban students possess as a result of participating in hip hop culture that can support full participation in science.

In fact, my experiences in one of the most essential activities of collectivity in hip hop, the rap cipher (where rappers/emcees gather to showcase their different skills) are most reminiscent of my experiences in a cell and molecular biology lab investigating cellular adhesion to extra-cellular matrix molecules. In both arenas, a close attention to detail, an ability to understand other participants' understandings, and a collective need to explain observations in ways that support a group's collective goals were necessary. In fact, I find that my experiences in both the rap cipher and the science laboratory are quite similar. As a former student in urban schools and presently as an urban science educator, I find that one of the main problems of urban science education is that the abilities of students who are a part of hip hop—abilities that are not focused on or given the opportunity to be expressed in the classroom—are precisely the attributes that need to be fostered in a laboratory scientist.

In this chapter, I construct a path to uncovering hip hop students' experiences that are integral to the teaching and learning of science. I achieve this goal by presenting an approach to engaging in a reality-based urban science pedagogy nested in my understandings and experiences as a participant in hip hop culture and science education (as student and teacher/researcher).

My experiences as a student in urban schools have provided me with some of my most valuable insights into the significance of the culture of hip hop to school and schooling. Through these experiences, my present standpoint is defined and expressed. This standpoint rests on my realization that the expression of hip hop culture in urban classrooms is a natural response to both the constricting structures of schools and the awareness that culture from fields outside of schools will always be transferred into the classroom. The expectation that a student's hip hop culture will completely stop being enacted once she reaches the doors of the classroom is impractical because a person's culture is deeply connected to who they are (see also Parsons, chapter 2, this volume). This realization could not be fully understood if I had not experienced this constriction of self in the classroom and dealt with the implications of being subdued within the classroom

In contrast to the classroom, the social fields where hip hop culture is most freely expressed allows multiple spaces for individuals to have voice. In these fields, the ability to fully participate in a particular social field is unlimited and, consequently, more complex ways of expressing one's culture are formulated. In these fields, different ways of rapping and varied ways of making beats become expressed. As a student in urban schools, I was able to witness events from a standpoint that my experiences as a science teacher would not have given me access to. However, my standpoint from each of these vantage points has allowed me to witness the profound impact that hip hop has for science pedagogy.

Hip Hop and Science Teaching: An Episode

Up until this point, I have discussed my stance as a participant in hip hop as a tool for understanding this culture, and how it can be used as a tool for science teaching and learning. This stance is articulated in order to argue for the fact that my standpoint as a participant in hip hop is one that presents me with an insider view from the perspective of both teacher and learner. However, through my work with urban students, who are currently deeply involved in hip hop, I learn new ways in which their culture affects their learning.

One example of this was in a high school physics class where I taught and conducted research. In this class, I noticed one of the high-achieving students murmuring to herself whenever I gave an assignment that required the class to solve questions that required calculation. This had occurred a couple of times in the past and caused me to be curious about what she was saying to herself. One day during class, I gave an assignment that required students to solve some questions on kinematics. As the class began solving the problems, Tasha (the student) began murmuring to herself. Not wanting to distract her from the assignment she was so deeply engaged in, I approached quietly and whispered to her, asking what she was saying.

Emdin: Tasha, what are you saying to yourself? You alright?
Tasha: (increasing her volume) Mic check one two what is this, *Tasha* is back to business. Ay!
Emdin: Jewelz Santana?
Tasha: That keeps me focused, keeps me going. I'm back to business. (turns back to her assignment)
Emdin: Do your thing, just be a little more quiet so we can get back to business too.
Tasha: (smiles) You got it E.

The lyric the student was saying to herself is from a rap song where the artist says "Mic check one two what is this, Santana's back to business." Tasha injected her name where the rap artist put his and used the line from the song as her way to stay focused in class. Through her recitation of that line, she had become so fully engaged in hip hop that the song not only served as a piece of music or way of expression but a focusing tool/motivation mechanism in the classroom. The lyric that Tasha recited helped her to reach a place where she was able to focus on answering the questions in the class, despite whatever else was going on around her. My experience with Tasha in this class stands in opposition to the ways that teachers described her in other classes, where they often complained about her "distracting and inappropriate behavior" and the fact that she needed to be in an alternative classroom setting. They did not realize that she was actually engaging in the classroom in very complex ways.

Despite the powerful impact of Tasha's use of hip hop on her learning and the profound influence it has on her and other students like her, many students walk into schools and are physically and emotionally forced to subdue their "selves" and "realities" in order for schools to maintain the existent structures that ultimately function to the detriment of student engagement in the classroom. This act of subduing oneself, which is forcing one to become a part of a foreign ideal that provokes discomfort at the risk of being even further ostracized from the school or the classroom, is shared by many students who are engaged in hip hop culture. In fact, the exchange I had with Tasha reminds me of one of my experiences as a student in an urban high school where my engagement in hip hop was a significant moment that marked the way that I view the hip hop students' experience in schools.

I was an eleventh grader and was sitting with some friends during the school lunch period. Napoleon banged on the lunch table with his fist and an afro pick and created the perfect drum beat. It was a deep bass generated by his fist and a light tapping snare that mimicked the drum pattern to an old school rap song. As the beat continued, Rodney started to freestyle (a rap verse made up on the spot). It was a slow, sluggish, yet distinct rap about his neighborhood and how great an emcee he was that seamlessly merged with the beat and even more seamlessly transitioned into the reggae-tinged rap of a Jamaican student. Before long, a small crowd had developed. We all stood, bobbing our heads to the beat and sounding collective noises in unison at the break in the beat when Napoleon rested his hands and the emcee simultaneously took the break in the beat to spit out a biting one liner. I waited for the rapper who had the floor to finish his verse and prepared to start mine. I gave him the customary cues that I was ready to start with more emphatic bobbing of my head to the beat and a repetitive sound after each of his lines. This ritual was the agreed upon (although never explicitly discussed) cue that it was time for the person who was rapping to hand the reins over to another rapper. As he wound down, I started with my introduction, eased into a steady flow of tongue twisters and handed the floor to the next rapper who amazed the crowd with his use of Haitian Creole in his verses.

At the height of the rap cipher, an announcement was made over the loudspeaker. "School safety report to exit 7, Student riot by exit 7." Seconds later, school safety officers came running towards the group of students and we quickly dispersed. I vividly remember grabbing my book-bag and running without knowing what I was running from or where I was running to. We knew that we were doing nothing wrong, but we also knew that our modes of expression, our thoughts and our ideas had no real place in the school and in response: we ran.

In the lunchroom that day, we all shared a common experience. As we rapped, about diverse subject matter, each verse that was shared articulated something different that was genuine to the collective group of participants. We rapped about our ancestry, our neighborhoods, our school, our dreams and our fantasies in a self-created social field with its own rules of engagement. In this sub-field of the school, we each had a voice that was valuable. We hung on each other's words as though they were the key to a new discovery, perhaps because we knew that we were in a space where we could share our opinions on anything from our classes to our lives.

In schools, and particularly in science classes, the feeling of participants in hip hop is that the voice and opinions of people who inhabit spaces far beyond those where the urban students dwell (completely removed from hip hop) are the only ones considered with regard to improving their experiences. We (participants in hip hop) provide deep commentary and insight into our schooling experiences that can be utilized to inform our experiences in schools but it seems as though no one has cared to listen.

Positioning Urban Youth in Science Education

In an age where Rudyard Kipling's six honest men (who, what, when, where, how and why—asked by scholars and researchers of urban education) have been repetitively summoned to seek answers to questions about improving urban science education, the answers they provide have neither effected much change in the attainment of educational outcomes nor improved the interest of students in urban schools. Ultimately, it has become increasingly evident that these honest men have become the foot soldiers of individuals and groups far removed from the people they are supposed to serve and do not have an accurate view of the lives of these people. Even when politicians, media outlets, or educational researchers have asked the right questions, they have not asked the right people about ways to improve their teaching and learning. In urban science classrooms, decisions about *who* gets taught science, *what* topics get taught, *when* certain topics are covered, *where* students learn science, *how* science gets taught, and in *what* ways students get taught, are removed from the very people in urban settings who the questions concern. Researchers who work in schools collect data and draw conclusions without the input of participants in the research. Consequently, the classroom becomes an extension of local and global fields beyond it; in these fields, the voice of certain populations as to who they are, what they stand for, and what it is that they need are often silenced.

In urban schools in the US, reports indicate that interest and performance in science continues to decline (NCES, 2006). Furthermore the influx of new immigrants into urban settings and the lack of provisions for addressing the needs of these populations cause those who focus on urban schools to identify that increased numbers of urban students are not successful in schools. In other words, if urban schools have traditionally underserved urban students in science, and these schools currently do not meet the needs of new populations who fill urban schools, an entire population of students in science (those of ethnicities that have been the primary population of urban schools like African Americans, and those who are new to these schools), are being underserved. This large population in urban settings consists of students with varying backgrounds, with many cultural barriers between their cultures and that of the school.

My work with these populations has uncovered that different ethnicities within the group named as urban students each report being disenfranchised from engagement in science within urban schools. These issues are not addressed in schools because of the informal instatement of a ban on student voice in classrooms and more covert ban on avenues that generate or create student voice like hip hop.

Populations from varying backgrounds are aware of the fact that schools are not addressing their needs in an effective manner and feel the effects of being unaccepted in schools. In response they are forging connections to each other under the banner of hip hop, where their collective oppression and position as other than the societal norm creates a space where their voice can be heard. This marginalization from the norm and collectivity under hip hop is a process that occurs not only within the United States or in urban areas. In fact, it is seen in places across the globe where one way of knowing and being has been established as the only way of knowing. In places as far away as South Africa and Germany, groups such as Black Noise and Advanced Chemistry (who inadvertently have names related to science) embraced a burgeoning hip hop movement in their respective nations by drawing young audiences and responding to the injustices in the social and political structures they experienced.

Within the process of forging identities under a collective, the process is not seamless. In fact, there are differences between people of different ethnicities in science classrooms and the struggles they have to overcome. However, I often find that those who have been deemed "other" by their singular (individual) experiences in schools and society often join together to respond to their oppression by finding a means/activity/process through which they can all respond to it. The realization sets in that there is a "we" as oppressed people that are being pushed away from having a voice or being successful, and that the various forms of hip hop (rap, dance, and graffiti art) provide an arena for voice.

My deep interest in hip hop partly stemmed from a feeling like I was not supposed to be an active participant in the science classroom despite my passion for the subject. My personal feelings of being ignored or treated as if I was not important within the classroom caused me to embrace the culture of those who were most pushed to the margins both outside of and within the classroom. My feelings, and those of individuals who experience this large-scale push to the margins, are those addressed in this book. This is also the population that hip hop artist Common refers to as *the people.* This population is comprised of those who are a part of hip hop from "Englewood [California] to a single 'hood in Botswana." These are also people who have been placed at the socio-economic and political periphery around the globe and who are expected to learn science in institutions that perpetuate this marginalization.

The focus on a science education from people for people stands as a shift from a top down (from the norm/expert to other/novice) approach to science education, instruction, and research. It focuses on the junctures where individuals who have been considered other than the norm merge to create a shared world around their oppression. The reality pedagogy that stems from science education from people for people works towards creating new spaces where value is ascribed to structures (physical, theoretical, ideological) that allow the learners of science in schools to have the agency to weigh in on the questions posed about improving urban science education for them.

For students from various racial, ethnic, and other identities, hip hop becomes a form of expression because expression or voice is strongly repressed by the structures of schools. This withholding of an avenue for voice is a form of robbery that removes students' agency in science classrooms. We have to focus on student voice as a key to providing students with the power to engage in the classroom (Furman & Calabrese Barton, 2006). However, the awareness that hip hop serves as an avenue through which this voice is enacted does not appear to be understood nor acknowledged.

When they are not provided with opportunities to express their voice, students in urban science classrooms feel like they are not part of the classroom. While they may physically be in the classroom, the ways that they describe the classroom often reveals that their selves are removed from it. When a person has no location in the world that properly is her own, she cannot participate in the relevant worldly affairs (Arendt, 1958). Consequently, the student who is immersed in hip hop will physically be present in the classroom and will not engage in a classroom with structures that do not support her interest. Hip hop culture is one that does support the chief interest of marginalized youth because it embraces those who are pushed to the margins and provides voice to these populations. In addition, its many hybridized forms make it a culture that participants from different backgrounds can collectively ascribe to. The particulars of different forms of hip hop are astounding and variations in language, style and subject matter run rampant. Regardless of this fact (the existence of different forms of hip hop), its articulation of the experience of the marginalized and provision of voice for this population remains paramount.

This shared experience/voice of the different participants in hip hop needs to be investigated to create tools that can be used to improve the teaching and learning of science in urban schools. The culture and the artifacts that are produced from this culture can be used to gain access to the voice of *the people* and create a pedagogy that focuses on the realities of those marginalized in science and education. I argue that their stance (as peripheral participants in society) is precisely why an exploration of their realities is important for improving an urban science education that is geared towards improving their educational experiences. By identifying their realities and using them as points for interrogation in teaching, science educators finally acknowledge that the identities of members of an oppressed group allows them to "see more, further, and better than the master precisely because of their marginalized and oppressed condition" (Smith, 2005, p. 8). The master here refers to the teacher who enacts a pedagogy that does not focus on student culture or backgrounds but on science work or the completion of tasks.

In urban science classrooms, the teacher and curriculum push "work" while failing to explore what the student needs to engage in science. The difference between work and engagement is necessary to elaborate upon because it is important to realize that teaching urban youth to work at tasks in a science classroom is simple and may be achieved by various means. As a student, I found that the presentation of subject matter in my science classes often revolved around topics and the scientists that developed certain ideas and concepts, but was never nested in scientists who were from backgrounds that were similar to mine or to how their scientific ideas were related to phenomena in the neighborhoods where I come from. Fostering engagement in school science and developing the thinking that supports an active involvement in it is a more difficult task that requires a focus on the realities of the marginalized.

The People Teaching the People

Teaching science to foster involvement/engagement in the discipline requires not only a deep awareness of both the subject matter and ways to engage marginalized youth. This process also requires an interrogation of the inhibitors to their engagement. In other words, if you want to teach me to become involved in the science classroom, you must first understand why I do not engage within it. Furthermore, you have to understand that I possess the skill set needed for understanding science and that I do express these skills in other fields. In fact, I may consciously make the choice not to apply this skill set in the classroom in response to the fact that my out-of-school experiences are not valued in the classroom. Most importantly, the realization is necessary that the urban students often have more insight into how they should be taught than their teachers have.

Such a realization comes with the awareness that the marginalized have knowledge on how to teach themselves that surpasses that which the power wielder has. In other words, as a part of hip hop, I have an insight or vision into my self that someone who is outside of my lifeworld and the lifeworlds I share with other members of my culture do not. This vision of teaching and learning that marginalized populations possess lacks visibility in science education. The lack of focus on this vision further maintains achievement gaps between participants in hip hop and those who are part of more mainstream culture. "Those empowered within dominant culture are visible, and visibility itself empowers" (Oliver, 2001, p. 108) and concurrently, invisibility and a blindness to the vision of participants in hip hop as to how to teach science, becomes disempowering. For example, a teacher attempting to start a biology laboratory lesson on genetics in an ethnically diverse urban classroom without accounting for students' knowledge in this area inherently supports a loss of voice for students whose voice is already often silenced because a genetic lesson in this class could benefit from students' input and provide students with an opportunity to teach and learn about each others' histories and how they relate to the scientific topic (see also Richardson Bruna & Lewis, chapter 6, this volume). In this instance the students' knowledges are not given a space to become expressed and the students become disempowered.

Disempowerment in one field is often the trigger for lack of interest in that field and a cause for hyper-engagement in another field that gives a space for the expression of agency. Lack of interest in science classrooms and the holistic engagement in hip hop is precisely a result of a loss of voice in the classrooms. For example, I became more infatuated with hip hop when it became a place where I could belong when I felt uncomfortable or uninvolved in the classroom. This infatuation grew into a deep devotion when the tenets of hip hop culture allowed me to express my feelings of frustration with not having a voice in school, with living in an overcrowded apartment at home or worrying about the well-being of my siblings. In hip hop, I had an avenue to display my intelligence or interests (however varied). I could write a rap with all the scientific vocabulary words I had taught myself and still be just as accepted as a friend who rapped about the violence in his neighborhood. The realities of the hip hop student, however varied, provide the deepest insight into their experiences. To teach these students, it is necessary to enact a pedagogy that is directly influenced by these realities.

Understanding a Reality Pedagogy

In describing a reality pedagogy for science education, it is necessary to describe the concepts of "reality" and "pedagogy" as distinct concepts that must converge. The amalgamation of these concepts is the first step to creating an approach to the teaching of science that benefits populations that have been removed from achievement or access and will be explained in detail in this section. Reality pedagogy acknowledges non-dominant standpoints and utilizes the position of those viewed as other as the point from which pedagogy is birthed, and from which, once developed, transformative teaching and research continues to feed. For example, the detailed exchange of batons in a rap cipher (as described earlier in my experience in the school lunchroom), the reading of cues as to when to actively participate in the cipher, the ability to make sense of complex oral text (speech, rhyme), are all attributes that some participants in hip hop possess which can guide the ways that teachers teach science or coordinate group activities in the classroom.

When a teacher mentions to the students that changes in the classroom will occur and that these changes are based on what she is learning or is willing to learn about the students' culture, it is a first step towards a valuing of student standpoint and a reordering of the hierarchy where teacher culture is positioned above student culture. Standpoint in this framework refers to "the social positioning of the subject of knowledge, [and] the knower and creator of knowledge" (Smith, 2005, p. 9). This process involves a positioning of the knowledge of the other (the hip hop student) as the norm. This re-positioning of standpoints in favor of the other is valuable in enacting a pedagogy that is responsive to the needs of the marginalized. Whereas engaging in this task may sound uncritically altruistic or, even worse, superficial (because of an age-old similar message about focusing on students' interests to improve science education), it is important to recognize that realities and interests vary and perceptions of realities are often misconstrued when they consist of speculation on the realities of urban youth by those who ascribe to the ontology and standpoint of the societal norm.

The rap cipher with my friends in the high school cafeteria was mistaken for a riot because the school administration viewed our meeting as a random gathering and not as an expression of our culture. As a group, we were viewed as being from a lower standpoint, which caused us to be viewed as rioters. The feeling that came with being ostracized from school and being a part of the group who joined together was our reality. The main idea here is that our subject position (as a part of hip hop) is directly related to our reality because each individual's reality is shaped by the position he or she is granted within society. With this awareness in place, reality pedagogy begins from the standpoint of a particular population in order to understand the affective (emotional) dimension of their experience and place value on it as a tool for teaching.

Understanding a Hip Hop Reality to Reach a Reality Pedagogy

For those who are part of hip hop culture, reality pedagogy is an avenue through which their realities are given voice and their potential to inform their own teaching and learning is released. "Hip hop reality" (like one's reality in general) denotes (a) the stance of each individual within each micro-world and (b) her experiences within them. However, it extends beyond any existent definition of a reality because it not only focuses on where a person stands within distinct social fields but also how she maneuvers through these fields based on being positioned as other than the norm, and how this maneuvering is expressed. My reality as a part of hip hop (as a student) in science classrooms involved how I constructed ways to maneuver through being dismissed from full participation in science because of my position in the world as part of hip hop. As I grew older, I had to learn how to subdue parts of my hip hop identity to become a fuller participant in science because there was no value ascribed to a hip hop identity in avenues where science is valued. Therefore, we can see that if one's reality is based on where one stands, and if this stance is in constant battle with constructions of its value by those with power on one hand, and one's belief about the true value of Self on the other, a hip hop reality can be seen as the struggle between who I am as a participant in hip hop and how I am perceived by others because I am a participant in Hip hop culture.

To explain how this reality is constructed and its influences on the science classroom, I draw from the notion of "double consciousness" and the articulation of the internal conflict in African Americans between being African and American (see also Parsons, chapter 2, this volume). In the science classroom, the conflict is between the students' position at the margins in the classroom and being central in an alternate world (as a part of hip hop). The specific focus on science here in this discussion is based on the fact that science in particular provides many more opportunities for participants in hip hop to participate if their culture is permitted to be a part of the classroom than other disciplines. This occurs because there is a specific science of hip hop and approach to teaching that can include phenomena explored in hip hop. Before this science of/in hip hop can be explored, it is necessary to provide a more vivid explanation of how and why the reality of the hip hop student is constructed. I do so in the following paragraph by providing the way my reality as a part of hip hop is constructed by my standpoints in different social fields.

My reality is a product of my ways of knowing and being and how I express them within each social field. This includes the self-confidence I feel when I am in my neighborhood, the feeling of being a full participant in the culture of my neighborhood, the way I dress, speak and interact and the nuances of the ways I interact with my peers (including everything from our head nods and handshakes) that are distinct in the ways we communicate with each other and makes me feel like a part of a hip hop community.

My reality is also a derivative of the established scripts about me by other people in fields beyond my present one. This includes their perception of the fact that I am a full participant in a culture other than their own, and the value (as less that the norm) of the way I dress, speak, and interact. My reality/standpoint is both situational and trans-situational (based on how they are enacted in each specific field but also based on other fields, how I am viewed within them and how I deal with them). For example, I view myself as a scientist, a scholar, and an educator, but must reconcile these identities with the fact that I am a young Black man who is voluntarily as well as involuntarily (a part of) hip hop. Voluntarily I choose to be a part of hip hop because it is an avenue that provides me with the agency I do not have in social fields like classrooms. I am painted with all the negative connotations placed on people who are from the hip hop generation. I may display a distinct identity as scholar, researcher, or scientist in a particular context, and I may be viewed as other than or less than an established norm in each of those identities because I am seen as a young, Black and hip hop before I am seen as scholar, researcher, or scientist. My reality is constituted by (a) the way that I navigate the positions where I have been placed, (b) my belief in the value of each of my various identities, and (c) the ways that I express how I navigate through my positions through way that I speak, dress, or move. I experience a marginalization based on my subject position as other than (and societally positioned as less than) a White older man who is a scientist or scholar. Despite my conscious decision to embrace my Blackness and hip hop nature, my voice to speak about my reality comes as much from my relegation as other (young Black male and hip hop) as it does from my voice as scientist, educator, and scholar. Consequently, I write through hip hop and express it as part of my culture and primary mode of expression while pushing for its efficacy as an educational tool both through my use of it as a writing tool and through my recommendation that it be used as a pedagogical tool.

Reality pedagogy is an approach to instruction that heeds the call for an attention to our true voice, an eye to the true picture of who and what we are, and attention to the ways that we navigate the existent flawed constructions of our realities. Through this tool, hip hop music, hip hop culture, the artifacts produced by hip hop, and students' perspectives about their experiences in schools are used as the primary tools for designing and implementing instruction. The nature of the rap cipher is utilized as a way to design group activities, class participation is nested in agreed-upon ways to exchange batons (as exists in a rap cipher), students are applauded and validated for utilizing new and innovative ways to present their ideas, there are equal turns at talk or participation, student experiences that cause them to become ostracized from school and schooling are addressed in the classroom, opportunities for an expression of cultural ideas and perspectives are allowed in the classroom, and concrete examples from hip hop are used in the classroom. At this point, the reader who can see the value of focusing on hip hop for the instruction of a hip hop populace would require a more thorough description of how this populace is created

The Creation of the Hip Hop Populace

Hip hop culture and the realities of those who are a part of this culture spring from both a voluntary and involuntary coalescing of those who are pushed to the margins. My experience being run out of the school cafeteria for participating in hip hop (described earlier in this chapter) was an experience that physically removed me from belonging within the school. However, it solidified my stance as a participant in hip hop. Because I had been chased away, I was involuntarily positioned as anti-school and pro-hip hop. This positioning occurred despite the fact that these two stances (anti-school or pro-hip hop) are not necessarily opposites. This either/or grouping of populations within the classroom where a hip hop population is found functions to make instruction or pedagogy less complex, because the instructor would only have to focus on meeting the needs of a small population.

The reader should envision a mass of individuals grouped in a central place and consistently moved out of the center based on their inability to meet particular requirements. For example, if a group of people gathered in a circle and were asked to take a step back based on being from a societally less desired race, socio-economic class, gender, etc., the individuals who would be furthest from the center would be involuntarily moved to those margins. The pedagogy engaged in the classroom would be directed towards the few who remain at the center. Those who are relegated to the furthest points outside of the circle are those who are a part of hip hop. Their commitment to hip hop and their full entrenchment in the culture occur in response to their experiences as part of the other (outside of the center or of the norm).

After being chased away from the school cafeteria for enacting hip hop culture in the lunchroom, the group of students who were in the rap cipher there met outside of the school and began to rap again. This time the topic of our rap was our disgust for the school safety officers and our hatred of a school that did not accept us for who we were. Our collective move away from centrality in the school caused our bonds of connection to each other to be stronger outside of the school.

The process I describe above (having individuals move further and further away from the center in one arena and fortifying an out-of-school identity in another arena) is not only evident in situations where students are interrupted from enacting their hip hop culture in obvious situations like the interruption of a gathering, but also in other more subtle ways. For example, I have experienced instances where the negative perceptions of participants in hip hop culture that comes from popular discourse (usually from news media and/or perceptions of hip hop that come from a glorification of its negative aspects) become the focus of teachers' perceptions of all students. I have witnessed the persistence of these negative stereotypes about students who are a part of hip hop in my experiences as both a high school student and teacher. As a student, I was asked by a teacher who consistently made comments about my attire how I was able to gain admission into the specialized high school I attended and become a teacher-researcher in urban schools; and I have conducted research with a teacher who asked how a student could be expected to learn with his hat over his eyes and his "oversized T-shirt."

These experiences, and many more like them, cause me to conclude that perceptions of hip hop participants created by media highlight and glorify the negative aspects of hip hop culture, and create a picture of a student who is not interested in schools who looks like a student immersed in hip hop. Consequently, teachers believe that students who dress or speak the same way as a highly publicized recently arrested rapper are not only viewed as unintelligent or prone to violence but also as abnormal (not the norm or not normal). In other words, we are viewed as though schools or schooling is not designed for us.

Articulating Pedagogy in the Context of a Reality Approach: Approaching a More Nuanced Science Education

If pedagogy—the science of teaching—gets enacted in ways that ascribe to a method of teaching designed for a population perceived as a norm, it also means that only certain groups of people are perceived as normal. The term "normals" describes the populations who are the societal norm and extends this idea of the norm to include being normal (Goffman, 1997). A transformative pedagogical approach challenges those who perceive themselves as normal (while others are viewed a stigmatized or less than normal) to reinvestigate their relegation of women, people of color, or any other classifications that vary from the established norm as having an inherent deficiency. If, for example, "normals" are the people who are traditionally engaged in science, and then we approach a reality pedagogy as a step towards reclassifying who is the "normal," it is necessary to see how science is culturally enacted within the classroom to further push those who could normally engaged in it to the periphery, or how it is used by those who are classified as other than normal. This process requires an approach to instruction that concurrently works towards removing the stigma associated with being a part of hip hop by showing that a normal valued science or scientific approach exists even within abnormal culture. Pedagogy then becomes a process that is not only concerned with subject matter delivery (as it is traditionally viewed) but as the point where an existing scholarly approach (in this example, science) meets with a new one and the hierarchies that are inherited by the positioning of approaches to education as normal or abnormal are dismantled.

In the case of science education, while engaging in the act of disseminating information about science to students, the teacher can redefine what is or is not science by placing value on ways of knowing or doing science that go beyond the norm. In enacting pedagogy that benefits students who are a part of hip hop culture, looking at hip hop's use of science is a necessary first step.

To enact this pedagogy and understand the use of science in hip hop, it is important to recognize that there is a difference between a real reality that is articulated by an "othered" population about themselves (this would include how participants in hip hop view their version of science) and a fabricated reality that is the perception of the marginalized population's reality from the viewpoint of those who are not a part of the culture. There are real forms of reality and fabricated ones: "the real, sincere, or honest performance; and the false one that thorough fabricators construct for us" (Goffman, 1963, p. 4). A real reality of the hip hop student would be the expression of a feeling of excitement, a search for opportunities to speak often and express knowledge of or interest in a topic. The fabricated reality (an outside person's interpretation of the feelings, goals or motives) of the same student would be the focus on the behaviors that are displayed as a result of the student's inquisitive nature, such as talking out of turn. For example, if a student describes how sound waves travel and uses the example of a shootout on her block as an example (as described by a student in a middle school science classroom I recently observed), and then spends a substantial part of the example describing the shootout in a hyper-engaged, loud, excited manner, I see her inquiry into the science of sound waves as being expressed in combination with the expression of her reality. In this instance, the teacher who taught this lesson saw the student's behavior as an attempt to derail the class and express inappropriate behavior.

The main issue that comes from this articulation of realities is whether or not science teachers can focus on the real reality of the student (which is the intent of the expressed behavior and occurs often through science), rather than the fabricated reality of who they see as "the hip hop student," which focuses explicitly on the mode of expression of the student and of the student's scientific thinking. The real reality is tapped into when the teacher/researcher focuses on the words/expressions of the hip hop student and the fact that their response to interest is more than often viscerally expressed because of the emotive nature of hip hop culture and the complex understandings of the student. This reality stands in a sharp contrast with the fabricated reality, which perceives students as difficult to teach, or hard to understand, based on factors like their hip hop attire or demeanor, and does not account for the ways in which they make sense of particular topics.

One avenue through which the real reality is reached is provided by the text of rap music. Smith (1999) articulates the value of looking at the text produced by a population as a way to achieve a layered insight into a reality that goes beyond a fabricated one. A look at a particular group's social life through text provides deep insight into the understandings of the population, as created text is often reflective and analytical. In the case of hip hop, much written text is produced by scholars of the culture. However, rap music, which is speech that stems from an artist's created text, is the most profound way to attain a phenomenological approach to the existence of a hip hop participant because the music provides both a viewing of the life of the marginalized, but also a hearing, feeling, and analysis of the oppressed population's standpoint, and an understanding of that population's notion of science.

Science in Hip Hop

The word "science" has as much of a storied history in hip hop culture as it does in dominant culture. In rap music (an artifact of, and one of the main elements of hip hop culture), when the word "science" is used (usually with the terms "dropping science," or "revolving around science," although not used as much in contemporary hip hop discourse), it demarcates the lines between a simple articulation of words and rhymes, and a descriptive, multilayered, and complex use of words that relays the ontologies of marginalized urban youth through the experiences of the performer. Science in this hip hop context has a root in the belief system of the 5-percenter Islamic sect and their use of the words "science" and "mathematics" as the method through which the logic behind real-life experiences and lessons about life should be learned by Black people in the United States.

The use of the words "science" and "mathematics" by the 5-percenters were absorbed by hip hop culture and became expressed in rap music. As a result, terms like "dropping science" in rap music, describe an act where there is a creation of expressions that are tied to the experiences of urban youth. This type of music is rooted in, although not necessarily driven by, its relationship to a religious origin, and used to explain the life experiences of the marginalized. This type of science is as complex as it is subjective and oftentimes describes how urban youth view themselves as the victims of traditional science. For example, Killah Priest of the rap group Wu Tang Clan describes in a song titled "Science Project" how the housing projects in New York City are a scientist's project. He then describes how people outside of the housing projects design what goes on within them as though some outside scientist is conducting an experiment. Concurrently, he engages in the activity of dropping science by discussing, describing, predicting, and analyzing a number of situations he has witnessed in a complex yet relatable manner. This display of subjectivity while meeting the criteria documented above (discussing, describing, predicting, and analyzing) is the expression of a skill set for both hip hop and canonical science.

I find the term "dropping science" one filled with much Socratic and literal irony as it describes traditional science being dropped (intentionally made to fall to a lower position), while it is referred to as a type of rap that is highly complex. This dropping of science concurrently elevates an alternative form of science that is found in hip hop. In other words, we concurrently drop science and create science. Furthermore, the perception of rap music as a superficial and categorically non-complex form of music causes general perceptions of the culture of hip hop as ignorant to be perpetuated. Rap artists who display ignorance/simplicity and produce uncritical music do exist and allow the critique of hip hop as non-complex to persist. However, the ignorance/simplicity/uncriticality of traditional science and education is exposed through rap that "drops science."

What Makes a Science the Science?

In canonical/school science, the inherent subjectivity behind what is presented as an objective discipline is hidden, despite the fact that it is subjectively in favor of the "normals." In other words, the roots of canonical science and its contemporary explications are just as rooted in emotion and the subjective as science in hip hop. The questions scientists ask, the limitations and boundaries of their experiments, the interpretations of these experiments, and the conclusions that are drawn from them are closely tied to scientists' culture and knowledge—which are based on their experiences as the norm. Furthermore, the religious overtones that guided scientific thought for centuries, be these Christian or Islamic, based on doctrine or belief, provided limitations regarding what scientific ideas were, and in some ways were allowed to be, viewed as science.

The key point here is that hip hop or rap that drops science not only shares a similarity to conventional science in how it is presented, but also holds a place similar to how the general population has arrived at science in general. By this, I mean that hip hop that "drops science" and individuals that "do science" share similarities because they are both grounded in standpoints. In addition, the average participant in each of their cultures holds them both in high esteem. The emcee/rap artist is as highly respected by participants in hip hop as the conventional scientist is respected in society in general. Therefore, when we as educators, researchers refer to scientists from this point on, we should include the participants in hip hop culture that produces a certain type of rap/knowledge/information and displays a similar skill to that of the traditional laboratory scientist or science teacher. With the similarities between a hip hop science and science in general having been articulated, I have provided a sufficient backdrop for an investigation of how science and science education can be made accessible to different populations.

One of the other major ties that binds scientists from both hip hop and scientists in more traditional settings is the belief that a scientist requires a certain acumen, special intelligence, or brilliance to become a full participant. However, it is at this critical point that the divergence between the cultures of hip hop and school reaches a point that excludes students embedded in hip hop culture from participating in science. Conventional science is presented as a discipline that is reserved for only a small group of people who make all the decisions for all the people. By the small group of people and the people, I refer to the traditional scientist (White male wearing a lab coat that almost all students draw when prompted to draw a picture of a scientist) and the people from marginalized populations respectively. By the people, I refer back to the populations referred to early in this chapter, and the large populations of students from various ethnic backgrounds that populate urban schools who are presented with information on science and scientists that shows that participation in the discipline is removed from them. To these groups, the "special ability" or intelligence to engage in science is presented as out of reach or unattainable. This is why these populations create a science through rap that "drops science." The creation of a new approach to science, or a new way of analyzing and understanding the world through rap, is a science that reshapes what conventional science is. This fact leads to the major difference between science as it is used in conventional discourse and science in hip hop discourse.

In hip hop, one of the major goals of engaging in science or creating/dropping science is that the science is just as relatable as it is complex. Despite the fact that the person who drops science is valued and possesses an ability that may not be enacted by everyone because it requires skill, the invitation to the listener or layperson is a key tenet. The emcee who does not have an audience is not effective, but the science teacher who does not reach the students (their realties) is still valued and respected. The divisions between these two types of science (hip hop and conventional) lies in the fact that school science is structured to not accept or support new ways of knowing or doing science, but rather focuses on an indoctrination of people into the culture of canonical science. Science or school in general is then viewed as a way to lose your credibility or position as a participant in the science of hip hop. A focus on the standpoint of those who are a part of hip hop culture provides insight into the fact that students engaged in hip hop possess many attributes (describing, explaining, predicting) that support success in science that are not fully explored in urban science classrooms because of the lack of value given to their culture.

Conclusions

Combining the reality of the other (particularly the realities of people who have been marginalized from attainment in science and/or education) with the pedagogical approaches necessary to reach these populations requires an understanding of their realities and a willingness to re-evaluate the pedagogical approaches currently in use in schools. Reality pedagogy in the sciences, from the standpoint of those who are a part of hip hop culture, requires a constant search for opportunities to manipulate existent structures for the benefit of these populations.

Those removed from hip hop attach the intellectual ability of the population to a seeming inability to attain success in a discipline like science; the idea that what is viewed as inability is a conscious effort to disengage from a marginalizing force is not focused on. I argue not only that this is the case, but also that a dismissal of the societally instated norm is a tenet of hip hop, just as a dismissal of hip hop culture as inherently anti-intellectual has proven to be a tenet of schools. While those invested in hip hop culture have been positioned outside of the norm in science classrooms, having to protect themselves from the perils of being on the margins of educational, political, economical, and societal centrality has provided these populations with an advantage in expressing where they are (articulating their stance). In addition, the taboo of being socially acceptable or conforming to a set of classroom norms does not apply because of their ability to rally around a banner of being collectively ostracized from the norm. Consequently a certain brutal honesty, unabashed realness, and insightful critique of self and other is developed and shared in a public medium like rap music.

This chapter has shared much of the theoretical basis for beginning the true work behind engaging in a reality pedagogy for marginalized populations. I hope to leave the reader with an approach for using the information provided in this chapter and engaging in the type of pedagogy recommended in work with students who are from the hip hop generation that I have found successful.

One of the major ways that hip hop culture can be introduced to teaching and learning in urban science classrooms is through the enactment of cogenerative dialogues. These dialogues are conversations with students and teachers about the science classroom and fields beyond it with the purpose of improving the classroom (Roth & Tobin, 2004). Whereas much work on cogenerative dialogue in science education has been implemented, I find that exploring student realities and standpoints as participants in hip hop culture prior to enacting the dialogues facilitates the process of gaining insight into students' lifeworlds. It is necessary to explore the descriptions of a reality pedagogy, the standpoint of the marginalized, and the meaning of science for the hip hop generation prior to having conversations with students, because they provide an insight into the conversations that it would not be possible to acquire during the conversations. In my cogenerative dialogues with students who are a part of hip hop culture and are students in science classrooms, my knowledge about their realities through the portal of my experiences as urban student of color in science classrooms provides me with a level of insight that assists me in restructuring the ways that I teach science. However, the realization that my experiences, a little over a decade ago in urban science classrooms only provide me with little insight in comparison to the new experiences of a new generation of immigrant students deeply committed to hip hop was necessary in order for me to enact a truly reality-based pedagogy. Our responsibility as educators and researchers working with marginalized populations is to introduce ourselves to approaches to teaching students of various backgrounds through the culture that binds them while engaging in a critical deconstruction of our roles as inhibitors of the teaching and learning of those who we intend to reach by our devaluing of their standpoints.

Coda

As a product of an urban public school system, I hold strong allegiances to teaching and learning in urban schools. My journey into higher education began with a fleeting interest in sound waves in an overcrowded general science class, and continues today as I research ways to hone the interests of students who express their interests in science classrooms. Stemming from my first attraction to the discipline, my educational trajectory in the natural and physical sciences has been influenced equally by educators who fed my curiosity and teachers who dismissed both my scientific interests and academic prowess. The lessons that I have learned through my experiences with these educators are the fodder for my exploration of this work. My deep passion for science in school was matched only by my passion for hip hop, and the dismissal of my interest in school science (from educators at different academic levels) was a cause for my passion and commitment to the science of hip hop. Realizing that my experiences (and those of my peers of different backgrounds) in schools were not unique when I engage in conversations with urban students in science classrooms today, I am convinced that transforming science education for the benefit of all students who share the wonder of physical and natural phenomena involves a thorough investigation of their culture as participants in hip hop.

References

Arendt, H. (1958). *The human condition.* Chicago: University of Chicago Press.

Furman, M., & Calabrese Barton, A. (2006). Capturing urban student voices in the creation of a science mini-documentary. *Journal of Research in Science Teaching, 43,* 667–694.

Goffman, E. (1963). *Behavior in public places.* New York: Free Press.

Goffman, E. (1997). Social life as drama. In C. Lemert, & A. Branaman (Eds.), *The Goffman reader* (pp. 95–107). Oxford: Blackwell.

National Center for Education Statistics. (NCES). (2006). *Nation's report card 2005 assessment results.* Washington, DC: U.S. Department of Education.

Oliver, K. (2001). *Witnessing: Beyond recognition.* Minneapolis: University of Minnesota Press.

Roth, W.-M., & Tobin, K. (2004). Co-generative dialoguing and metaloguing: Reflexivity of processes and genres. *Forum: Qualitative Social Research, 5*(3). URL: http://www.qualitative-research.net/fqs-texte/3-04/04-3-7-e.htm. (Retrieved October 21, 2007.)

Smith, D. E. (1999). *Writing the social: Critique, theory, and investigations.* Toronto: University of Toronto Press.

Smith, D. E. (2005). *Institutional ethnography: A sociology for people.* Lanham, MD: AltaMira Press.

6 Sister City, Sister Science

Science Education for Sustainable Living and Learning in the New Borderlands

Katherine Richardson Bruna and Hannah Lewis

On the wall of the downtown post office in Ames, Iowa, a mural offers a surprising message to patrons: the lives of Iowans and of the Mexican newcomers immigrating to Iowa are inextricably linked. The mural depicts, on the left-hand side, an Aztec farmer bending over to work the land. There is a pyramid in the background, as well as a statue of a god. On the right-hand side, the mural depicts a parallel scene but with an Iowa farmer instead. The pyramid has been replaced with a factory, the god with a microscope. Dividing the scenes is an ear of corn (Figure 6.1). When Mexican students in a neighboring demographically transitioning community are asked to consider what the mural's message is, they don't miss a beat. "*Se trata de la unificación del hombre sobre la cultivación del maíz,*" they answer. "It's about the unification of man over the cultivation of corn."

Given the ethnic- and class-based segmentation in their new Iowa "host" community and, more particularly, the accompanying marginalization these Mexican newcomer students experience from peers and teachers in school, it is difficult to reconcile the harmonious idealism of this statement with the harmful realism of their daily lives. Indeed, corn *is* a unifying factor in the histories of Iowa and Mexico—it has played a crucial, even iconic (as the mural attests), role in both economies. This essential site of similarity, however, is lost amidst a whirl of anti-immigrant sentiment that produces a falsely absolute Us (Iowans)/Them (Mexicans) difference in the New Borderlands communities in

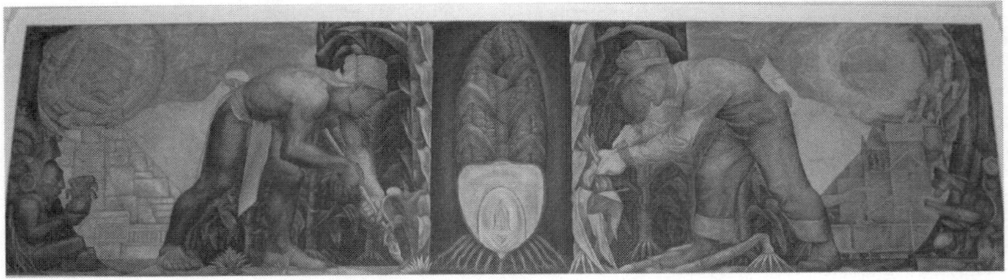

Figure 6.1 A mural on a post-office wall in Ames, Iowa depicting the Mexican–Iowan corn connection.

which they settle to work, communities which, through their presence, they revitalize at an astounding pace. What might be gained, we ask, by attending to the gap between the students' idealized response of human universality and the very particularized reality of their dehumanization as immigrant labor? And, more importantly, what does any of this have to do with science?

This chapter, as do others in this book, provides an alternate framing of what counts as and in science. We believe that the "Mexicanization" of the Midwest poses a compelling challenge to conventional understandings of the purpose of science education, even "reform-based" varieties that take responsive science instruction as their stated point of departure. Without legal documentation nor the assurance of remaining permanently in the US (nor the desire to), the futures of many Mexican newcomer youth lie outside the normative parameters of science professionalization that guide science instruction in this country—to secure the nation's scientific edge and, with it, its economic prosperity and political advantage. Most, with their families, hope to return to Mexico, but if they stay, as "illegal aliens," they will be, in most states, barred from receiving essential financial assistance, making the prospect of getting into the science "pipeline" impossible to achieve.

This popular metaphor of the pipeline, used to describe culturally and linguistically relevant recruitment and retention efforts aimed at increasing the numbers of under-represented youth pursuing science or science-related fields, is, for many Mexican youth, nothing but a pipe dream. Despite its nod to inclusion, this metaphor is generated out of a "theoretically impoverished and politically visionless" agenda of science education (Kyle, 2001, p. xi). This agenda:

> ignores the wider complex of socio-cultural and political factors that influence the ways schooling is structured to benefit some students at the expense of others . . . neglects the fundamental issues of the place of science in any larger social context and fails to acknowledge the political situatedness of science . . . and invokes a narrow image of science by valuing the technical interests of the empirical-analytical sciences at the expense of the practical interests of the hermeneutic-interpretive sciences and the emancipatory interests of the critical sciences. (p. xii)

Science education can mean something more than a personnel project aimed at generating a highly skilled workforce to help the United States keep pace with global markets. It can be a place where students inquire into the processes of the natural world, discover human interactions with those processes, and explore the implications of human understandings of these natural processes, and the interactions and activities to which they lead (the knowledge project known as science), on human experience and natural life more broadly. Simply put, science can be, instead of mere economics, a matter of person formation, of learning more about oneself as a human-natural being and about the world in which one lives. It can be an endeavor unbounded by political or national borders and untarnished by the competitive rhetoric that is capitalism's calling card.

National borders, after all, in a globalizing economy become necessarily more diffuse, problematizing any conceptualization of science education that involves the production of the national citizen and the growing of the national economy. Citizenship and economics, in globalization, are distributed *across* national borders. Thus, for science education to be relevant in the transnational communities generated from globalizing movements of people, it must itself adopt a distributive approach. Students' resources for science learning will be distributed across both their, in this case, U.S. and Mexican contexts, as will be their sites of application for that science learning in their lives in and beyond school. This is the reality to which we call upon science education most urgently to respond if it is to acquire any substantive meaningfulness in the borderland lives of U.S.-Mexican youth.

This chapter is written from, and demands science to be inclusive of, the standpoint of this borderland, transnational (Mexico *and* U.S.) lived experience. This standpoint defies, we argue, the abstracted and generalized forms of social organization—unitary national and political belonging—used in science discourse to talk about educational processes and outcomes related to Mexican youth. Instead, we call on the local and particular actualities of the Mexican experience, actualities that the abstractions and generalizations of a bounded and binary (U.S. *or* Mexico) nationality fail to capture. It is the "everydayness" of transnational identity and, most particularly, of the "work knowledges" (Smith, 2005) related to and generated out of Mexico–U.S. migration that we urge science educators, in light of present demographic trends, to squarely consider.

Our discussion here radicalizes even further the notion of a "citizen science," a populist understanding of the concerns, interests, and activities of science, because we extend its meaning to those people *most* on the margin or "border," as it were—those who may not be, in fact, "citizens" at all. So, in accomplishing our larger goal in this chapter, of providing an alternate framing of what counts as and in science, we provide here a description of a border science pedagogy that fills in what Kyle indicates is currently missing in the broader science education agenda: attention to learners most disenfranchised by science instruction; awareness of the socio-political context in which the practice of science is embedded; and allowance for other forms of science knowing. We ground our border science approach in a tale of two "sister cities" (one in Mexico, one in Iowa) deeply affected by the forces of globalization, specifically the growing corporate control of food and agricultural systems. In telling this tale, we profile Mexican farmers (some in Mexico, some in Iowa) who are trying to withstand such forces as they struggle to hold on to traditional work knowledges and sustain their agrarian lifestyles. Out of this telling, we believe, a clear need and path for a border science pedagogy in the demographically transitioning Midwest emerges, one that would not only achieve Kyle's critical science education goals for *all* students, but begin, as we argue, to heal a community as well. Our tale begins with the story of corn.

Twin Cradles of Corn: Mexico and Iowa

Corn does indeed, through its cultivation, preparation, and consumption, link the historic consciousness of the Aztecs in Mexico some 8,000 years ago to that of its modern human beneficiaries throughout the globe. Methods of corn's domestication spread quickly throughout the Americas, and, with European colonialism, throughout Europe, Asia, Africa, and China as well. Today's increased-yield hybrid varieties feed the livestock upon which most U.S. households depend for their protein, and sweeten the confections and soft drinks that line supermarket shelves. If we are what we eat, we are, then, "processed corn, walking" (Pollan, 2006, p. 23).

Interestingly, after their "English Learner Science" course, the same Mexican newcomers who offered their interpretation of the post office mural join with others crowding the cafeteria to buy a Coke. They owe their enjoyment of this beverage both to their ancient ancestors and to modern agricultural scientists who introduced hybrid corn to the farmer. The juxtaposed figures of the mural represent the lineage, as it were, of corn care: one figure reveres corn as god and the other regards corn as commodity. The latter, in particular, and, importantly, assisted by the tools of modern science, dramatically increased corn's yield and spurred on the corn-derived processed foods industry that is responsible for today's typical American diet.

But, returning to the tension we discussed at the beginning of this chapter, as much as corn has been a uniting force in human consciousness, it has also been divisive. European settlement across the US and the resultant extermination and displacement of native peoples was a push for acreage on which, among other things, to grow corn. Green Revolution programs aimed, in the middle of the 20th century, to stave off poverty and malnutrition resulted, instead, in continued disparity. Peasant farmers in countries like Mexico, where agricultural scientists sowed the new corn "miracle seed" with the promise that tens of millions of extra tons could be harvested, could not afford the large amounts of fertilizers and pesticides the new corn varieties required. Only the big growers could afford the expense for these "inputs," and, additionally, these big growers grew to make profit from export rather than from feeding their own citizens.

These historical forces in Mexico were in an interplay with those in the US in significant ways. In the US, two economic crises, one after World War I and another after the Vietnam War, deflated land prices, and with them, the profitability of corn. At the same time, corn hybridization and other technological innovations enabled farmers to produce more per acre, with less labor. As overall farm output increased, prices declined, and farmers needed to farm more land to maintain their level of income—they had to "get big or get out." Farms big enough to stay in the game grew a narrow selection of subsidized crops on land previously dedicated to more diverse rotations, including pasture, hay, and small grains. In Iowa, farmers placed their bets on corn, which grew well in the region's fertile, rain-fed soils, and received government subsidy support. But even large-scale farmers were subject to global market fluctuations, and to terms of trade dictated by agribusiness, a rapidly consolidating industry of input suppliers and farm output processors. Abundant, cheap corn was a boon not only to makers of high-fructose corn syrup and other corn derivatives, but also to the livestock industry. It enabled large-scale producers to raise livestock in concentrated animal feeding operations (CAFOs) on contract with large, highly mechanized packing plants. In such a context, small family farmers in Iowa and throughout the Midwest hardly stood a chance. Like their counterparts in rural Mexico, they found themselves competing with industrial farm giants, and losing.

For Mexico, the situation became even more grave when, as part of their entrance into the North Atlantic Free Trade Agreement (NAFTA), the Mexican government seized communally held farmland in the name of industrialization and privatization and eliminated their corn subsidies. Cheap corn from the US flooded across the "free trade" border, further squeezing out the peasant farmer. He, in consequence, in order to feed his family, headed north to slaughter and pack the hogs that had been displaced off Midwest farmland and corralled into CAFOs, all in the quest for corn.

In sum: "Industrial agriculture has not produced more food. It has destroyed diverse sources of food, and it has stolen food from other species to bring larger quantities of specific commodities to the market" (Shiva, 2000, p. 12). To this we would add, more specifically, it has forced Mexican and U.S. farmers off their land and, in so doing, begun to decenter generations-old work knowledges developed in and around the intimacy of the agrarian lifestyle. For Mexicans, this has meant taking up new, mechanized, menial, and alienated activity in, most prominently, the meatpacking houses of the U.S. Midwest. Because of its industrialization and its effects on the ecosystem, including human activity, "corn [in human hands] is the protocapitalist plant" (Pollan, 2006, p. 26).

Read in this way, the history of "the unification of man over the cultivation of corn" is one of a shared struggle of the small farmer against corporate interests. With industrialization, the connection between the eater, the land that grows what is eaten, and the people that tend that land has been lost. We, in "developed" countries, have become industrialized eaters. As industrialized eaters, most of us do not see that immigration from Mexico, so vilified in the media, is at its core about the politics, not of culture nor language, but of food. As the Mexican newcomer students leave the cafeteria and mill around the front of the high school, they sit at tables on which someone, probably someone eating a sandwich containing processed meat and drinking a corn-syrup soda, has angrily scrawled the words "Fuck Mexicans" or "Mexicans Go Home." Mexican students are victimized in this way by Iowan students who are themselves victims of a culture of industrialized eating and blinded to the history that unites, and divides them, over food production and consumption.

Mexican Work Knowledges and Sustainable Agrarian Lifestyles in the Twin Corn Cradles

Views and Voices from Villachuato

Villachuato, Michoacán, Mexico, because of its meatpacking-based migration, is the unofficial "sister city" of Marshalltown, Iowa. Home of Swift and Company, the third largest hog plant in the world, Marshalltown has been a growing destination for Villachuatans since 1989. While there were only 248 Hispanics in Marshalltown in 1990, the 2000 census figure was 3,265, or 12.6% of the entire population. When survey results specifically from the Mexican community in Marshalltown revealed that approximately 3,000 were from Villachuato, it meant, in fact, that the number of Villachuatans living in Marshalltown was actually larger than that of those residing in Villachuato itself.

"Está muy difícil la vida. Aquí no hay trabajo./Life is very difficult. There is no work here."

Migration to the US is one strategy rural Mexican farmers use to adapt to the economic crisis caused by the protocapitalization of corn. To shield against loss, family members leave their communities for periods as part of an overall family strategy permitting them, ironically, to remain on their land. Unable to support their families through farming, given the price of renting land and buying seeds and necessary "inputs" such as fertilizer, pesticides, herbicides, and water (access to irrigation has, since 1972, been in the hands of federal authorities), younger men from Villachuato go north to work at the plant in Marshalltown. The *remesas* (remittances) of family members in Marshalltown are key to survival in Villachuato. Without them, as María Dolores, a 71-year-old Villachuatan mother of seven children, six of whom are working in the US—five at the plant in Marshalltown—described it, "*La gente no tiene más que pa' mal comer.*/The people only have enough for bad eating."

The experience of "bad eating" is part of the historical community context of poverty in Villachuato. María Dolores speaks to this poverty when she says of her deceased farmer father:

> *Mi pobre padre en el puro cerro con azadón para sembrar frijól y maíz, y traía lena para vender en un burro que tenía. Amanecía uno sin que comer. En el suelo, había gente que sí dormía en el suelo. Todo se mojaba de los techos. Andaba la gente con los pies a raíz. Si no tenía uno ropita, pues se lavaba uno y se la volvía a poner.*/My poor father [worked] all the time in the field with a large hoe to plant beans and corn, and he carried wood to sell on the donkey he had. Day would dawn without any food to eat. On the dirt floor, people would sleep on the dirt floor. The roofs leaked. People walked in

rags. If you didn't have a change of clothes, you washed them and then just put them right back on.

It is only because of migration to Marshalltown (among other U.S. destinations) that families in Villachuato are now able to build bigger homes with solid roofs and real floors, to have changes of clothes, and to not wake up hungry. Work in the community that would make these outcomes possible simply doesn't exist. As María Dolores explains, "*Está muy difícil la vida. Aquí no hay trabajo. Que tengan su trabajo diario que pueden trabajar, no.*/Life is very difficult. Here there is no work. It's not the case that one has their daily work that they can do." While some families still plant, there are not many that do. "*Poca gente pone.*/Few people plant," she laments.

However, from appearances, agriculture seems to be alive and well in Villachuato. Walking down the streets, you note the pungent waft of a family pig lazing outside someone's house in town. Horses, cows, bulls, and goats munch cut straw or hay in small, enclosed dirt yards. One boy playing in the streets clutches a tired fighter rooster, while others, walking and on horseback, herd goats through the streets (Figure 6.2). Tractor implements sit parked in front of houses. Pick-up trucks drive by loaded with wheat, sacks of cut straw, melons or corn stalks. Men drive John Deere tractors into driveways. You can't walk more than five blocks before passing a seed and fertilizer supply shop (Figure 6.3). A butcher's shop sits right off the main plaza (Figure 6.4).

Figure 6.2 Boys in Villachuato tend to farm animals as part of their daily responsibilities.

Figure 6.3 Seed stores like this one illustrate the community's agrarian identity and also the pervasive presence of American (DeKalb) hybrid corn varieties.

Figure 6.4 This is a butcher stand off the plaza, temporarily vacated by the teenage girl who runs it for her family.

These visual signs belie the fact that agriculture is for most a losing proposition in Villachuato. As María Dolores says:

> *Mira de las tierras que siembra aquí, siembran los que tienen con que sembrarlas, que tienen tractor y todo eso, pues les queda algo, pero los que no tienen con que sembrarla, no les queda más que puro para pagar, luego este los que prestan dinero pues ellos sacan para sembrar, y nada más se van en pagar y se queda la gente igual.*/Look, about the land that people cultivate here, those who can plant with whatever they have to plant, those who have a tractor and all that, well for them they might make something, but for those who don't have anything to plant, all that's left for them is to pay, those who borrow money, who take it out to plant, they have to pay and they're in the same situation they were in when they started.

To assist rural families, the Mexican government gives *dispensas* or food rations. María Dolores describes the inadequacy of what she receives:

> *Nada más un paquetito de harina y un litro de aceite. Primero daban un kilo de azúcar, ahora dan medo kilo, y una bolsita de lentejas, una bolsita de frijól, pero el frijól no sirve, no se coce. Pues dan lo que no sirve. El arróz no sirve. Está duro, está feo.*/Nothing more than a little package of flour and a liter of oil. First they gave a kilo of sugar, now half a kilo, and a little bag of lentils, a little bag of beans, but the beans don't work, they don't cook. So they give us something that doesn't work. The rice doesn't work. It's hard, it's bad.

To have more than just enough for "bad eating," María Dolores and her 80-year-old husband, Víctor, are completely dependent on the *remesas* they receive from their children in Marshalltown. They have their home, six cows, and a bit of land that lies fallow. Víctor spends his days sitting in a sling-back chair he has placed in a small patch of corn growing in the backyard of their home. A proud man, he describes himself now as *un pájaro con boca abierta*, a bird with its mouth open, waiting to be fed.

"Mi mamá me trajo a este mundo yo creo con el fin de crier animales./ My mother brought me into this world I believe with the purpose of taking care of animals."

Disabled from years of hard work as a *bracero* (farm laborer) in the asparagus fields of California and from his life as a *jinete* (horseman) in Mexico, Víctor resists his children's suggestion that he should sell his animals and his land. Agriculture is his identity. He says: "*Mi mamá me trajo a este mundo yo creo con el fin de crier animales porque a mí me gusta mucho el ganado y los caballos y todo porque así me crie.*/My mother brought me into this world I believe with the purpose of taking care of animals because I like livestock and horses and everything because that is how I was raised."

Víctor's neighbor, Oscar, is a younger man approaching the age of 50. Having once headed north to work in the meatpacking plants, Oscar is now committed to trying to make his living as a farmer in Villachuato. Behind his small adobe house are four little huts for four sows. He sells their piglets to a broker, who comes to town periodically calling out up and down the streets for people to sell him pigs. In addition, Oscar sometimes sells pigs to local butchers to be consumed locally. Chickens and turkeys scratch around his yard and a horse grazes on the corn stubble in a small field beside his house. On fields just a five-minute drive from his home he grows strawberries and a few rows of cucumbers to sell, as well as the sorghum he needs to feed his animals.

Despite being surrounded by the crops and livestock he raises, Oscar worries about what he's going to feed his family tomorrow. Unlike María Dolores and Víctor, who have a new brick house and are otherwise supported (from large appliances, like refrigerators, to small amenities, like toilet paper) by family up north, nobody in Oscar's family works in the US. He blames the Mexican government for abandoning the countryside and argues that farmers must be supported:

> *No progresa uno, pues nada, no vale pues lo que siembra. Una tonelada de sal te cuesta 2,000, una tonelada de ore te cuesta 4,000, un bultito de semillas de unos 19 kilos te cuesta 1,280, asi que pues, si inviertes mucho dinero aquí, pero para el tiempo que tu cosechas, apenas lo puedes sacar y tu trabajo, si tú lo metes el trabajo, no ganas ni 5 pesos por día. ¿Sí, me entiendes? . . . ¿Trabajo los seis meses para producir el maíz y sin ganarle un cinco.*/One doesn't make any progress, nothing, what you plant isn't worth anything. A ton of salt costs you 2,000 pesos, a ton of sea salt costs you 4,000 pesos, a handful of seeds of 19 kilos costs you 1,280, in this way you invest a lot of money, but with the time it takes for you to harvest, whatever you can get and your labor, you don't earn even 5 cents a day. Do you understand? I work six months to produce corn without making a cent.

On the role that international economic policy plays in construing his desperate situation, Oscar is an astute critic:

> *Nosotros no podemos competir con los Estados Unidos. Estados Unidos tiene mucha economia, a los agricultures el gobierno siempre los ha ayudado. Aquí no. Aquí ¿cómo vas a competir? Meten mucho maíz también para allá, y sacan maíz de aquí, pero esos son los ricos, los ricos son los que están haciendo ese negocio, no nosotros, nosotros producimos, y a nosotros no compran a un precio barato. Hay grandes compradores pues que sí tienen dinero y esos lo venden a un precio más bien allá en los Estados Unidos. Por decir nosotros que nos pusiéramos y ¿quién nos apoya, o qué? Por decir que al libre comercio pasarlo, pues ¿para allá cómo? Nosotros no tenemos la manera.*/We can't compete with the United States. The United States has a lot of economy, the government has always helped the farmers. Not here. How are you

going to compete? They send a lot of corn there and they take it away from here, but these are the rich people, the rich are those that are doing this business, not us, we produce, and they buy from us at a cheap price. There are big buyers that have money and they sell at a higher price in the United States and who can help us out? They say that it's about free trade but over there, how can it be? We don't have the means.

Yet amidst these troubles and the distrust they cause, many people in Villachuato do continue to farm in some way at some level. Farming is tradition. Farming, as Víctor explained, is what they were raised to do. The roots of agriculture run deep and long in Villachuato; and this is reflected in the work knowledges that, despite disruptions and interruptions in family transmission posed by migration, are being passed down to the younger generations. Children acquire the rhythms, sensibilities, and skills related to agriculture from their relatives, importantly, through apprenticeship-like relationships that provide them with early and ongoing exposure to the body of information and related practices and understandings they are expected to someday master. As Lewis wrote in field notes from her visit to Villachuato:

Yesterday, I awoke early to drive out with Eduardo [a Villachuato farmer] and his 18-year-old grandson, Jorge [visiting, in fact, from Marshalltown], to the fields. It was raining so they didn't use their John Deere tractor or plant their ready bags of brightly colored fungicide-treated seeds. Instead, we went out to the corral to feed 20 or so meat and dairy cows. Eduardo and his grandson divided the work of emptying a mixture of chopped straw and grain into several small feed troughs. They noted that a few cows were pregnant and none of the others were lactating or needing to be milked . . . Then, today, I went with Ricardo [María Dolores' and Víctor's only remaining child in Villachuato] to his fields, where he was loading bags of cut straw into the back of his pickup to take back and store near the corral down the street from his house. His six-year-old son Juan lingered in the field with the men as they slung heavy bags in the air, and returned with them to help pack the bags into storage. I learned that Juan also accompanied his father in the mornings out to milk the cows, and sometimes out of town on days when Ricardo walked a dairy herd to graze on grassy strips between fields. Juan was learning how to take care of animals so he could step in for his father when needed. After milking, Ricardo and Juan bring the buckets of warm milk into the kitchen, where the women in the family divide it into pint-sized pails to sell to neighbor children. In both these cases, Jorge and Juan are acquiring a knowledge of systems, of how things fit together—how animal feed (grain, cut straw, fresh growing grass) is processed into food for humans (milk). At the heart of such knowledge is awareness of processes that reveal connections between living things.

"El número uno es la formación de una persona./The number one thing is the formation of a person."

Drawing on the work knowledges that children acquire through their authentic and meaningful agriculture-centered experiences, some schools in rural Mexico have, in fact, an *agropecuario* (farming) focus. For example, in La Escuela Secundaria Técnica de Angamacutiro (Angamacutiro Technical Middle School), located right outside of Villachuato, all curriculum is linked to the core activity of growing crops and raising livestock. Classrooms look out over fields of corn that the students themselves tend and the stockades that hold the livestock that the students themselves feed (with the crops they grow), raise (including artificial insemination and castration), kill, and, ultimately, butcher. Then they use the meat for lunches at the school and sell the rest. In the words of the school's *director* (principal), Fernando Andrés Rojas Cortés, at the heart of the agricultural focus is the goal of *autosuficiencia* or self-sufficiency because this is the larger goal of the rural community. But this goal goes beyond just having the knowledge to feed oneself; it is about understanding, as he says, *el por qué* (the why) behind such practices.

At the middle school in Angamacutiro, students learn in ways that connect to, affirm, and extend their everyday work knowledges as they strive to achieve self-sufficiency towards, in fact, self-actualization. By their graduation year, students are expected to be critically analytic and reflexive thinkers. They develop these skills through regular *discursos* (lecture and discussion groups) in which they are asked to directly confront the challenges that face their agricultural communities, among these poverty, crime, drug addiction, immigration. As Fernando explains:

> *El número uno es la formación de una persona y para formar a una persona no estamos trabajando con una máquina para que digan cuánto es 2 × 2, cuánto es 1 × 1 o que diga letras. Eso no es una persona. Ese es el proposito acá: formar personas que cuando lleguen al tercer año aquí con nosotros tengan una habilidad crítica y que sean analíticos críticos y reflexivos, que tengan reflexiones de su actuar.*/The number one thing is the formation of the person and in order to form a person we are not working with them as machines so that they can say how much is 2 × 2 or 1 × 1 or so they can say letters. That is not a person. This is the objective here: to form persons so that when they arrive at their third year with us they have critical skills and are critically analytic and reflexive, that they are reflexive on their actions.

The food system in and around Villachuato is, then, subject to the competing desires of maintaining, even in the face of grave social and economic challenges, traditional self-reliant agricultural practices and of abandoning those practices (in part or whole) for a life lived *allá* (there—the north). The futures of Villachuato's youth will revolve around the question of what is the decision they can, quite literally, live with. Each decision brings its own consequences, and, if the decision is made to cross *sin papeles* (without papers), those consequences can be traumatic, if not fatal. One consequence, if the crossing succeeds, is the adoption of a new work-related identity in the different food systems context of Marshalltown. In this new context, the recognized work knowledges of rural Mexicans are radically and fundamentally transformed.

Views and Voices from Marshalltown

Making the crossing fundamentally changes two interrelated features of rural Mexican identity: production and consumption. The familiar (meaning both "recognizable" and "associated with the family") land-based ways of working and eating, which are at the heart of an agrarian identity, come to be replaced by labor and living that is alien (different) and alienated (disassociated). Given how deeply ingrained a farming sensibility is in the lives of Villachuatans, it is difficult to conceive what it is like to trade in the knowledge and skills one has about working the land for the repetitive de-skilled work on a mechanized meatpacking line, or the preference one has for eating fresh from the land with the reality of "big box" retail. For Mexican newcomers, striving for an agrarian lifestyle in the rural Midwest environment of Marshalltown means making space, around the edges of plant work, for traditional agricultural practice. Here, agriculture becomes not so much oriented around immediate economic sufficiency (although with mouths to feed back in Mexico that is always a priority) but around cultural maintenance.

"Ahorita están matando 9,300 puercos y no completamos las ocho horas./ Now we're killing 9,300 pigs and we don't even work eight hours."

When María Dolores and Víctor's eldest daughter, Angélica, came to Marshalltown in 1991, she was, she recalls, the only Mexican woman, and one of few Mexicans, in general, on the line; today, she emphasizes, there are only two Americans. The Mexican work force has increased the kill speed at the plant tremendously:

> *Mataban 7,500 puercos en ocho horas. Ahorita están matando 9,300 puercos y no completamos las ocho horas. Salimos 15, 20 minutos antes de las tres de la tarde.*/They used to kill 7,500 pigs in eight hours. Now they are killing 9,300 and we don't complete eight hours. We leave 15 to 20 minutes before 3 o'clock in the afternoon.

The majority of Mexicans in Marshalltown are, as is Angélica, simply laborers in a production process owned and operated by someone else (Swift and Company). The holistic knowledge of how to select, plant, and harvest seed, or, more closely, to butcher a whole animal, is replaced by isolated, repetitive action within the confines of the plant's grey, windowless walls (Figure 6.5). Although most production jobs in meatpacking require little training, the work is physically demanding and difficult. It involves standing for long periods, lifting

Figure 6.5 A view from the near-empty parking lot of the Swift plant in Marshalltown, IA, taken on May 1, 2006, when immigrants nationwide showed their support for immigration reform by boycotting their jobs.

heavy objects, moving on slippery surfaces, and using dangerous cutting and grinding tools and machines. Animal slaughtering ranks among the U.S. industries with the highest rates of injury and, because the largely undocumented status of meatpacking plant employees militates against workers advocating on their behalf, the actual frequency and intensity of injuries will never be known.

The Mexicans laboring on the line are also consumers, and this activity, with their move to Marshalltown, is also radically transformed. Ironically, they are among those who buy the produce that "big box" stores such as Wal-Mart purchase at a cheap price from large Mexican growers and sell at a grand profit, the ones, who as Oscar from Villachuato pointed out, are helping to squeeze out the small Mexican farmer. The Marshalltown Wal-Mart, for example, has become a Super Wal-Mart due to the growing Mexican presence and profits. And it now carries items Mexican families in Marshalltown used to have to go out of their way to get—beans, chillies, and *ojas y masa* (corn husk and meal) for *tamales.* As Angélica explains:

> *La Wal-Mart es super center. La Kmart es super. He visto que este pueblito ha progresado mucho porque todo tanta gente immigrate vino a vivir.*/Wal-Mart

> is a Super Center. The Kmart is Super. I have seen that this little town has progressed much because so many immigrant people have come to live.

"Si no tuviera animals, no sería quien soy./If I didn't have animals, I wouldn't be who I am."

Arturo, who describes his job at the plant as "*mucho trabajo, poco sueldo*" (much work, little pay), wants to be a full-time farmer. As a child, his family members in Mexico were poor small farmers. They had a few goats, some chickens and pigs for home consumption, a few dozen sheep, and a small field of corn, beans and squash. It was Arturo's responsibility to walk the herd out of town and back each day to graze. The fondness he developed for his animals and his awareness that he learned something from being in their presence is clear in his words:

> *Yo era él que las cuidaba borregos o chivos. Yo era él que andaba siempre entre de ellos. Es algo bonito porque de ahí agarra uno. Se aprende uno mucho. Se da uno cuenta. En mi casa, yo tenía que andar cuidándolos diario, diario, diario.*/It was I who took care of the sheep and goats. It was I who always herded them. It's a lovely thing because from this, one grasps. One learns a lot. One is aware of a lot. In my home, I had to walk with them [the sheep and goats as they went out of town to graze] every single day.

Arturo then left home at the age of 13 to work at his brother's butcher shop where he learned to butcher goats, sheep, cows, and pigs:

> *Se aprendía destrozar una vaca, a matar, desde matar hasta descuartizarla todo. Me enseñé a lavar tripas, me enseñé a pelar patas, a pelar cabeza, a sacar pieza por pieza de las vacas—de las partes de la vaca.*/I learned how to slaughter a cow; from slaughter to cutting it into pieces, everything. I taught myself to wash the intestines, to skin the feet, skin the head, to take the cows apart piece by piece—the different parts, the parts of the cow.

Arturo also learned valuable animal husbandry skills, including how to castrate boars, wean piglets, and to assist sows in difficult labor. Today, in Iowa, with legal residency status, he has a herd of a few dozen doe goats, flocks of chickens and turkeys, and a vegetable garden for home consumption. He gains supplemental income to his work at the plant by selling his animals and charging an additional fee to cook *barbacoa* (barbecue). Most importantly, through his farming activity, Arturo provides fresh, quality meat for his family and keeps alive, also for his family, the knowledge and skills he inherited from his rural Mexican upbringing. About the centrality of agriculture to his identity, Arturo says, "*Si no tuviera animals, no sería quien soy . . . Desde chico . . . me gustan los animales. Yo no puedo estar sin tener animales.*/If I didn't have animals, I wouldn't be as I am . . . Since I was a child, I have liked animals. I can't exist without animals."

Arturo is pursuing farming in the rural Midwest not for its profit potential, but for its quality-of-life and quality-of-work benefits. He is increasing his family's overall well-being by, though minimally, augmenting their income, more strongly, by securing access to preferable food, and, most substantially, by carrying on in the New Borderlands the knowledges, skills, and traditions of his rural Mexican upbringing. Because of the value Arturo places on his agrarian identity, he has made space for the work knowledges associated with it around the margins of his otherwise industrialized identity.

Views and Voices from the Sister Cities

These views from the sister cities of Villachuato and Marshalltown help us understand how Mexican migration is enmeshed with economic and cultural strategies of agrarian sustainability. Migration north allows the family members left behind to continue to engage in the, if not economically viable, at least symbolically vital, agrarian practices of their ancestors. And, for the (im)migrants themselves, making space in their lives for agrarian practices generates a sense of well-being through revitalizing traditional work knowledges and identities valued in the home-place that they have physically, but not emotionally, left behind. The maintenance of these agrarian activities, despite their displaced context, shields Mexicans in the New Borderlands against the physical and mental stress of their industrialized identities, identities they acquire through their alien and alienated—dehumanizing—labor on the line.

Because of the rigid ethnic and socio-economic segmentation in the Marshalltown community, however, the profile of work knowledges that Arturo possesses and the fact that he is actively working to transmit those to his children are lost. He becomes just another Mexican at the plant. The butchering skills Arturo learned at his brother's shop, for example, are made invisible by the mechanized and de-skilled labor he performs on the line. The December 2006 Swift raids that drew national attention to the demographic transition in the Midwest and underscored the extent of the undocumented labor on which the meatpacking plants rely, formed an indelible image of Mexicans as criminalized workers that serves to even further one-dimensionalize Arturo's presence and ignore the strengths he and his children bring to the community and schools. Without direct attempts to tap into these work knowledges through curriculum and instruction, the education of Mexican youth will be as though they are, as Director Fernando at the school in Angamacutiro warned, simply machines. "No estamos trabajando con una máquina para que digan cuánto es 2 × 2, cuánto es 1 × 1./We're not working with them as machines so that they say how much is 2 × 2, how much is 1 × 1," he insisted. But with the value of Mexicans measured, as Angélica showed us, in kills per hour, the mechanization of their education—not as people to form but line labor to learn—doesn't seem too far behind.

Towards a Border Science Pedagogy of and for the People

For science education, the implication of this automatic equation of Mexicans with meatpacking wage work is captured in an ethnographic account from the same meatpacking town (Richardson Bruna & Vann, 2007). In an "English Learner Science" classroom in the rural Midwest, the teacher prepares her Mexican newcomer students for a pig dissection by framing the activity as preparation for work at the plant (Figure 6.6). She says:

> The animal that we are going to dissect . . . are pigs, because all of you guys keep saying. Oh I can't wait to go down to Bensen [the hog plant], so I can make money. Well if you're gonna learn about a pig before you go to work at Bensen, we're gonna start talking about a pig. And we're going to start talking about what is inside of a pig. (2007, p. 31)

Not only does this statement make explicit the assumption that Mexican students are the future labor force for the plant, it also highlights the teacher's lack of awareness of the work knowledges these students bring into the science classroom. These students, as Arturo's profile as a butcher's apprentice makes clear, come from families that already know a lot about pigs. Indeed, in Richardson Bruna and Vann's account, we see how a particular student, Augusto, demonstrates his extensive knowledge of pigs—how to calculate their

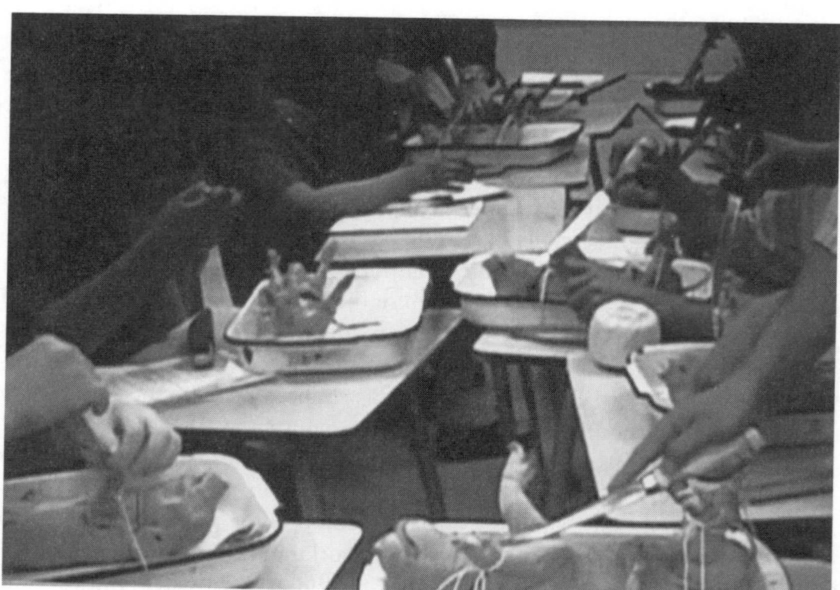

Figure 6.6 Mexican students in an "English Learner Science" classroom dissect pigs so they can, according to their teacher, "go to work at Bensen [the meatpacking plant]."

gestation period, what size they are when they are born, and what miscarried pig fetuses look and smell like. And, strikingly, throughout the lesson, we continue to see the teacher return to her framing of them as de-skilled laborers at the plant, despite Augusto's assertions to the contrary. In a perversion of the schools-as-servants-of their-communities model we saw in Fernando's Escuela Secundaria Técnica de Angamacutiro, Mexican students in the New Borderlands are socialized to accept their roles as wage-laborers in a segmented social and economic context. The goal here is not self-actualization, and critical analysis and reflection, but self-erasure and submission and subjugation to the status quo.

We assert, returning to Kyle's critique of science education that opened the chapter, that the privileging, in modern science, of the theoretical (the realm of words and ideas celebrated by an empirical-analytic framework) over the practical (the realm of actions more characteristic of a hermeneutic-interpretive framework) is an essential feature in making possible the invisibility of Mexican work knowledges in schools and communities. For this reason, the actions and engaged discernment-of-meaning activities of generations of ancient Aztec "farmers" as they tended corn crop after corn crop for millennia on end can be overshadowed by the achievements of a few white "agricultural scientists" in the last half of the 20th century. Favorable mutations that the Aztecs achieved that increased ear size, for example, mutations attained through a process of learning from (not "about") the plant they regarded as a god, are deemed trivial and sentimental, not "scientific." And it is not just the work knowledges of the Aztecs and their Mexican descendants that are demeaned in this way. Scientific elitism is an equal opportunity enterprise. Wallace and Brown (1988), the self-titled "fathers" of hybrid corn, place under a photograph of White "dirt farmer" George Krug the following caption: [He] "did great things with corn but lacked words to describe his ideas" (p. 75). This caption speaks to the contempt for manual labor, in general, that underlies the development of modern science.

In reality, however, the role of the manual laborer in the development of science must not be underestimated. Midwives and mariners, shepherdesses and shipbuilders each accumulated the factual data that constituted the fundamental building blocks of scientific knowledge. They were the forebears of empirical observation, experimentation, and causal research. But "scientists" took up and codified the "tradesmen's" knowledge, and then deprecated it, depreciated it, and sold it to someone else—and here's the clincher—for whom, in a budding capitalist economy, the once-specialist became a wage-worker (Conner, 2005). This rise of science—the substitution of a god with a microscope, of a pyramid with a factory, of the Aztec "dirt farmer" with a White "agricultural scientist"—is the story-behind-the-story told by the post office mural. It is this story that enables the non-recognition of the agrarian work knowledges that Mexican, and, in fact, many White students and their families, bring to Iowa schools and communities.

Since one good story deserves another, we aim to provide a counterstory to this grand narrative of science and, in addition, to the illusion of separateness that characterizes the social segmentation of demographically transitioning communities like Marshalltown. (On counterstories, see also Calabrese Barton, chapter 8, this volume.) It is a counterstory we envision as enacted through the implementation of a border science pedagogy that re-privileges traditional work knowledges associated with small farming and, in so doing, re-centers the shared connection of human relationship with the land. Our impulse as counterstory tellers rests on this belief: if Iowan students and the Mexican students who share their schools and classrooms were provided with opportunities to co-inquire into the transnational interdependence of their families and communities, they would discover common features of their histories and begin to build bridges where pathways across difference are desperately needed.

One such feature and potential bridge is the challenge of agrarian sustainability. Today, in fact, three-quarters of farm households in the US must generate outside income in order to keep their farms. This reality fundamentally changes their work and family life identities. "I loved seeing things grow, being my own boss and working together as a family," a small U.S. farmer says (Fussell, 1992, p. 160). These sentiments clearly resonate with the feelings expressed by Víctor and Oscar, and those of many other Mexican farmers who not only can't *work* together as a family, with industrialized agriculture, but also can't *live* together as a family. This is where the Iowan and Mexican small farmer experience differs: the Mexican farmer suffers not only economic and psychological displacement, but, of course, coming to the U.S. Midwest, radical geographic displacement as well. Providing White students with educational experiences that afford them a shared vantage point with their Mexican peers—in this case, the common platform of the struggling family farm—builds a foundation of identification from which empathy can flow. Comprehending, with compassion, the causes of such radical Mexican geographic displacement is essential, we believe, in countering the hostile anti-immigrant sentiment that characterizes many demographically transitioning communities in the Midwest.

In this way, with the goal of revealing the shared disenfranchisement of small Mexican and Iowan farmers at the hands of corporate agribusiness, and, further, of investigating the role played by science and "agricultural scientists" in enabling this disenfranchisement through the deprecation and depreciation of the "dirt farmer's" work knowledges, we hope to construct a classroom counterstory of corn (see also Brandt, chapter 3, this volume). We conceive of a border science unit that does this by making sister city relationships born out of globalization, such as that between Villachuato and Marshalltown, a matter of pedagogical, not just passing, importance. In those relationships lies the potential of forging connections between White and Mexican students and families in the New Borderlands communities, and of reclaiming science as a project of and for the people. An undeniable additionally desirable outcome is that of elevating the status of Mexican newcomers in the New Borderlands science classroom and closing the gap between the ideal of human universality glimpsed in, for example, their response to the post office mural and the particular reality of their dehumanization, in these New Borderlands contexts, as merely line labor.

We understand the unit we propose as "bordered" in the sense that it is responsive to the transforming social contexts that result from globalization, and, more specifically, from the movement of peoples across geographic and political borders. Such border crossing construes contexts for which the conventional nation-state model of being and belonging, and resultant approaches to teaching and learning, are no longer descriptively nor instructionally adequate. With populations whose lived experiences, both Mexican and White alike, are configured through the cultural, linguistic, and economic practices that exist not *in* a particular nation-state place, but *between* nation-state places, pedagogical practices must, we assert, be reflective of and responsive to these emerging realities. Thus, a border pedagogy is necessarily a pedagogy of and for the people because it attends to the people's own bordered lifeworlds. A border *science* pedagogy, more specifically, re-centers bodies of knowledge-generating practices associated with the bordered lifeworlds of globalizing communities by regarding them as valuable meaning-making mechanisms. And it does so with a commitment to enhancing human life using these generative knowledges and practices. The crisis of industrial farming and eating that unifies and divides Mexican and Iowan families in the New Borderlands is one that, we believe, a border science pedagogy of and for the people can help address.

"Who Owns the Seeds?"

"Who Owns the Seeds?" is an interdisciplinary (science, social studies, language arts) border science unit intended for middle school implementation. While we conceive of the unit as being taught in both Marshalltown and Villachuato, we have used the U.S. national science standards to guide our design and intend the transnational science teaching and learning opportunity afforded by the Marshalltown context as the principal location for implementation because, as we have explained, an important objective is to break down the walls that exist between Iowan and Mexican students in that community. For this reason, the unit is meant to be taught in an integrated classroom in which Mexican students learn alongside their White peers. This is, in fact, the integrated context in which science instruction now takes place in Marshalltown. Given the emphasis in No Child Left Behind educational policy on "highly qualified" teachers, Mexican newcomers no longer experience a transition year where they receive content instruction, like science, from ESL teachers. They are, instead, assigned to mainstream science courses, regardless of their level of English proficiency, where whatever language support they need is provided through the services of a bilingual instructional assistant. Thus we note that effective implementation of the unit we propose would involve thoughtful and thorough use of English Language Development strategies on the part of the science teacher and her/his strategic collaboration with the bilingual aide. In this regard, we take up and embrace the understanding that curriculum should be "a complicated conversation across diverse perspectives" (Sleeter, 2005, p. 116) and, in this case, diverse voices.

Because at the heart of the curricular conversation we aspire to spark is the goal of generating transformative intellectual knowledge among students, we showcase, in the unit, an individual whose lifework exemplifies such knowledge. Vandana Shiva is a physicist and advocate for ecological justice who speaks for the small Indian farmer and other historically subjugated agrarian communities. In so doing, she challenges dominant discourses about food production. We have chosen modified texts by Shiva that appear in the rethinking schools publication, *Rethinking Globalization: Teaching for Justice in an Unjust World.* The articles "Stealing nature's harvest" and "Relocalization, not globalization" will supplement the reading students will be doing from the traditional science textbook (Padilla, 2001).

The unit in this textbook around which we have designed our border science exploration is "Seed plants." We chose this unit as our point of departure because of immediate conceptual links between our plant of interest, corn, which is a seed plant, as well as a textbook support (a unit-concluding feature on the conversion of plants to chemicals) that takes corn as its specific focus. Given Iowa's leading role in the development of alternative corn-based fuel sources, we particularly felt this support would provide students with one important avenue for another meaningful, real-world connection between the science content in their textbook and contemporary social issues.

Whereas we reserve a role for the traditional textbook in this way, our unit relies, on another avenue for meaningful, real-world connections: the students' own lives. After students are drawn "into" the unit with the basic and broad leading question "What is a seed plant?," they will be invited to bring in and discuss examples of seeds from seed plants in and around their homes. This will launch a corn-seed-growing classroom activity that will be the backdrop of inquiry as they move "through" an exploration of the history of the cultivation and hybridization of corn, the Green Revolution, and the rise of agribusiness and globalization. Students will be asked to gather information related to those topics by collecting oral histories of family and community members involved in farming. At this point in the unit, the collective focus narrows in consideration of the family- and community-specific nature of the inquiry. Because the classroom will, necessarily, by virtue of the Marshalltown population, consist of Iowan and Mexican students, these histories will, in sum, provide and elicit a transnational perspective and analysis of the kinds of work knowledges present in those sister cities related to agrarian life. These histories will then form the basis for the broadening out of inquiry, in the next phase of the unit, to an examination of sustainability and biodiversity more generally. Here the students arrive at the unit's essential question: Who owns the seeds? They will prepare presentations in response that synthesize all elements of their unit learning.

The capstone event will be a student-hosted open house in which families and community members are invited to view and discuss student presentations that illustrate their own acquired transformative intellectual knowledges. This event is a critical one as it not only provides an opportunity for authentic assessment of student content knowledge but, moreover, promotes dialogue between constituents of the Marshalltown educational community that rarely, if ever, are invited to come together over substantive issues of mutual concern. This activity of discerning the mutuality of their experiences is a crucial outcome of the border science unit, one that will continue for students as they take their learning "beyond" the classroom and school by exchanging projects with peers in Villachuato (perhaps their desk mates next year?) who, as we envision it, would have been engaged in a similar path of inquiry. An overview of the general scope and sequence of this border science unit, and its anchors in the national science standards, is depicted in Figure 6.7.

We have designed this border science unit because we believe that curriculum must be responsive to the context in which it is taught. It is not just ineffective, but irresponsible, we assert, to teach a science unit involving, in this case, corn and corn-related technologies without acknowledging how the very history (and history-laden tensions) related to those issues is living and breathing in

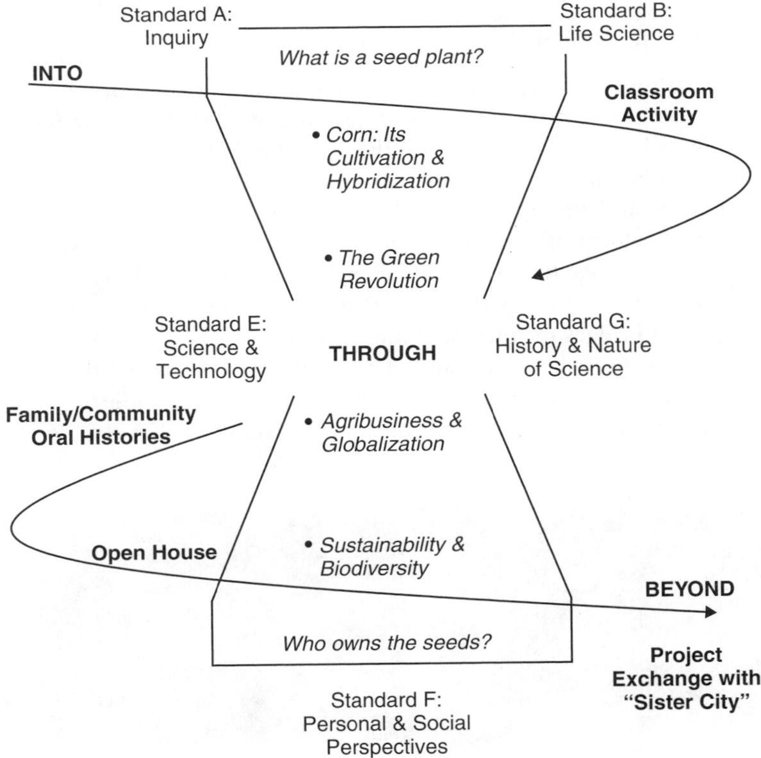

Figure 6.7 A Border Science Curriculum Unit on Seed Plants.

front of you. In envisioning this border science unit, we dream of a classroom where the business-as-usual social segregation of Iowan and Mexican students begins to be replaced by partnerships rooted in authentic science inquiry that positions all students' lives as critical curriculum. Indeed, corn has been, as astute Mexican newcomer students observed, a site of the "unification of man through cultivation"; it has also been a site of long and ongoing human struggle.

A science of and for the people is a science that engages in, not ignores, this "complicated conversation across diverse perspectives" and, in so doing, opens up the individual to new understandings of their connection to not just science, but also to school, society, self, and, significantly, the "Other." Mexicans come to Marshalltown because they have, in essence, "lost the farm." This is an experience all-too-familiar to rural families in the Midwest. By leveraging this undeniable point of connection, we can convert the Us (Iowans)/Them (Mexicans) mentality into a We mentality, a We who recognize the mutuality of their place of wounding in the history of corn, a history usurped by corporate greed that has crippled connections to the land and stolen the very seeds of a shared humanity. We can, through a new kind of science, tell a new kind of story.

Schooling for the Global Citizen

On the wall behind a fast-food joint in Marshalltown, Iowa, a mural offers a surprising message to patrons: the lives of Iowans and of Mexican newcomers immigrating to Iowa are inextricably linked (Figure 6.8). This teacher-and-student-created mural is a comforting sight in a New Borderlands community characterized by anti-immigrant sentiment, yet experience suggests that the kids here who wield paintbrushes in peace now, declaring that they are "*una gente*/one people," wage turf battles of belonging later. In this climate, it will take concerted effort on the part of all stakeholders in the educational community to build bridges of understanding and begin to heal the dis-ease of mutual suspicion and self-distortion that exists among Iowans and their new Mexican neighbors. Science educators, we argue, have their own role to play in addressing the borders that exist in the New Borderlands.

Figure 6.8 A mural in Marshalltown, IA, celebrating its connection to Mexico.

A border science pedagogy, like that we have outlined here, fills in what is currently missing in the broader science education agenda: attention to learners most disenfranchised by science instruction; awareness of the socio-political context in which the practice of science is embedded; and allowance for other forms of science knowing. In so doing, it takes as its point of departure that the goal of science learning is person formation, not worker preparation. Just as movements to reclaim food democracy insist that the kind of living associated with industrial eating is not sustainable—both from a health-of-the-individual and a health-of-the-planet-perspective—movements to make science education more responsive to cultural and linguistic diversity must insist that the reason for doing so reaches beyond technological innovation and economic gain towards a more sustainable vision of learning. The nation-state model of competitive advancement upon which such rhetoric rests makes no sense in transnational communities like Marshalltown. Whose economy, after all, "owns" the work of Mexicans, like Angélica, who regularly sends money home and sustains her family in Villachuato through the strategy of migration? Whose education system "owns" the learning of Mexican children who, as part of this migration strategy, follow their families back and forth across national contexts?

More specifically, what does "preparing workers for a global economy" mean in Marshalltown anyway? For many Mexicans, it means preparation for their labor on the line while, for many Whites, it means learning the "cross-cultural competencies" to be managers of those line workers. This is not a sustainable educational vision for learning nor for living. Many Mexican youth will be alienated, as they indeed are, in Marshalltown, by curricular and instructional practices that do not reflect their reality. As a result they will drop out, as they indeed do, to pursue the only opportunity they can see—that offered by Swift & Company—and thus reproduce the rigid ethnic- and class-based segmentation of the community.

Mexicans, yes, are dehumanized on the line but the experience does not start, nor end, on the line alone. As we write, accounts come in about Immigration and Customs Enforcement (ICE) conducting home invasions. ICE has been waiting outside of schools, we are told, in hopes of apprehending undocumented parents while they pick up their children. Such children, we hear, are asking their principals "Is there a kids' immigration police?," "Will they come and get me at school?," and, more painfully still, "Would you let them?" Between the noise of the hog slaughter and the silence of a well-enjoyed pork meal, these children are trying to make their voices heard. "Is this a safe place for me?" they are asking. They are owed an answer. Science educators are among those who need to reply.

A global economy demands an educational model of global citizenship in which challenges arising from globalization become points of inquiry, not ignorance, in the curriculum. We need our own curricular model of *el por qué* (the why) and its goal of educating critically analytic and reflexive thinkers. If we are to have an economy of border-crossing products, producers, and production, we must provide an education, unbounded by national and political boundaries, that can travel. We invite science education professionals to take up such a model in the New Borderlands of today and tomorrow. It falls to us all, in the end, to help sow new seeds.

Acknowledgments

We gratefully acknowledge the insight and enthusiasm of Ellen Fairchild in helping us conceptualize a border science unit that would speak to the experiences of Villachuatan youth and their families living in Marshalltown. We hope our sharing of Villachuato's story in some small way begins to repay the generous hospitality of that community which has, on different occasions, extended a warm welcome to us all.

References

Conner, C. D. (2005). *A people's history of science: Miners, midwives, and "low mechanics."* New York: Nation Books.

Fussell, B. (1992). *The story of corn.* New York: Alfred A. Knopf.

Kyle, W. C., Jr. (2001). Foreword: Toward a political philosophy of science education. In A. C. Barton, & M. D. Osborne (Eds.), *Teaching science in diverse settings: Marginalized discourses and classroom practice* (pp. xi–xvii). New York: Peter Lang.

Padilla, M. J. (2001). *Life science.* Upper Saddle River, NJ: Prentice Hall.

Pollan, M. (2006). *The omnivore's dilemma: A natural history of four meals.* New York: Penguin.

Richardson Bruna, K., & Vann, R. (2007). On pigs and packers: Radically contextualizing a practice of science with Mexican immigrant students. *Cultural Studies of Science Education, 2,* 19–59.

Shiva, V. (2000). *Stolen harvest: The hijacking of the global food supply.* Cambridge, MA: South End Press.

Sleeter, C. E. (2005). *Un-standardizing curriculum: Multicultural teaching in the standards-based classroom.* New York: Teachers College Press.

Smith, D. E. (2005). *Institutional ethnography: A sociology for people.* Lanham, MA: AltaMira Press.

Wallace, H. A., & Brown, W. L. (1988). *Corn and its early fathers* (rev. ed.). Ames: Iowa State University Press.

7 Cultural Encounters, Countering Enculturation

Metalogues about Cultures and School Science

Carol B. Brandt, Christopher Emdin, SungWon Hwang, Eileen Carlton Parsons, Katherine Richardson Bruna, and Wolff-Michael Roth

In the five preceding chapters, we meet a range of people whose everyday knowing somehow comes into conflict with official science and science education. There are Ms. Herlton and LaShaundra, two African American women of very different generations, who exhibit a lot of everyday knowing about the natural world but whose local everyday knowledge is made to pale in relation to official scientific knowledge. There are Anselmo Quam and his Zuni people, who have developed tremendous agricultural knowledge that allowed them to farm in an area where the industrialized ways of the only apparently globalized and globalizing agro-industry no longer work and yet who are encouraged to abandon their ways in favor of the tractors and plows of the Whites. There are SungWon and her research participant (Miko), who are confronted with themselves as they experience the radical changes that come with transnational migration, which can be viewed as an analogy for the "migration" from everyday life into the science classroom. There is Chris Emdin, who comes to understand the needs of his students through shared experiences in the hip hop culture and its ways of relating to others and its knowledge, both of which exhibit similarities and differences. And, finally, there are Arturo, Augusto, and the other Mexican immigrants and migrant workers, whose agricultural knowledge is in danger of being obliterated in the schools and towns of the American Midwest.

In all these chapters, we can feel the inner contradictions in science education, between the teaching of canonical science knowledge, which requires abandoning—at least in the classroom—the everyday knowledge that has allowed all these people to be successful in their daily lives, and the grounding of students' learning in their prior experiences, which students have to abandon and abrogate in the face of science that becomes a colonial and colonizing endeavor. In the following four metalogues, we take on issues of authenticity, the critique of our concept of "culturing knowledge," the relation between academics and practitioners of science education, place-based education, cultural appropriateness of science education, and the critique of science and its epistemology.

*

Michael: One of the criticisms that academics often face is that what they are writing about does not concern, affect, or import to school practice. One of the reviewers of the original proposal "firmly believes that we need to bridge theory and practice and extend our thinking beyond 'culturing knowledges' to envisioning what that would look like in diverse pre-college classrooms across the country." But s/he then suggests that we are academics who "are at it again: writing for themselves, making important points based on significant research but . . . not truly impacting teachers' practice and the significant classroom events that define contemporary science education." In their chapter, Katherine and Hannah already outline a curriculum that crosses the boundaries between current official science curricula and the lives of the people they are working with. What are some of the ways in which our chapters can mediate what teachers and students do in contemporary science education?

Katherine: I see two important ideas in this reviewer's comments that need to be unpacked. One is the notion that the work of "culturing knowledges" does not *already* exist in diverse classrooms; the other is that this work does not impact teachers' practice *unless* we academics make it do so. My research on the experiences of Mexican newcomer adolescents in science instruction has convinced me that we need to reconceptualize teaching and learning to bring students' home- and community-based understandings from the margin to the center because, for these students and for all students, that is their point of reference. To talk as if the work of "culturing knowledges" is not already taking place in classrooms is to privilege a decontextualized mode of believing, behaving, and being that I doubt really exists at all in practice, so the question is less about envisioning what this work looks like but moving toward research and teaching that makes the work of "culturing knowledges" part of normative understandings of science education. Similarly, because students are already doing the work of "culturing knowledges," teachers' practices are already impacted. It is our work as academics to understand precisely this: how does students' "culturing knowledges" work intersect with and influence the work of mainstream science pedagogy to generate hybridity and innovation thereby enriching science teaching, learning, knowing, and doing?

Chris: It is important for us to recognize that the criticisms that are fielded against academics about how accessible our work is to teachers and students is based on the way that we present ourselves and our work to the general public. While some assumptions are founded on general assumptions about the insularity of the academic community that are beyond our control, many more of the critiques we receive are supported by the activities that we engage in, the ways that we enact research and the ways we choose to disseminate the outcomes our work. I do acknowledge that the constellation of responsibilities that encompass the life of an academic and the priorities associated with each of them force the maintenance of a culture that focuses on maintaining a scholarly presence in the academic community. However, the choice to continue to produce scholarly work that is removed from students and teachers even while we discuss them in our work is a decision that is made by the individual scholar.

Carol: I've been really struck by how strongly the personal narrative within the interview can throw wide open the ways in which we situate ourselves in science. Eileen's interviews with Ms. Herlton and LaShaundra provided a location for Eileen to challenge the boundaries of science and re-think authenticity in science. My own experience with interviewing the farmers and gardeners at Zuni was a watershed experience for me, one that allowed me to shift my standpoint. There is something about the intense listening to another person's narrative that I think should be incorporated into our educational practice of science. I'm wondering if this is what you were getting at in your chapter, SungWon, and Michael, when you were talking about how embodied passivity where one is open to a different approach?

SungWon: Your point is interesting. The frequent occurrences of boundary-crossing in contemporary science education (e.g., between everyday practice and school and workplace) exemplify the power to act as a central aspect of literacy. The aspects of passivity in (scientific) literacy indicate not only the capacity to become competent in using diverse cultural (scientific) resources—which is often emphasized by the term "multicultural"—but also the power to draw on cultural resources, and thereby expand control over life conditions and the room to maneuver for the Self. In teaching and learning science, this comes with the continuous translation of the science curriculum and this is apparently different from the standardized (teleological) schooling that puts the value on speaking the standardized language and conducts evaluation from it. I think the concept of institutional ethnography and the chapters in this book show the performative, collective, affective nature of scientific literacy and the possibility to develop educational practice for it.

Michael: But some critics hold this against us, the fact that we interview people to *understand* what they do and how they learn rather than engaging in the teaching. Rather than viewing our mutual endeavors as contributing to the same problem, practitioners of teaching and (academic) practitioners of research are often put into two camps.

Chris: The fact that academic culture primarily functions to maintain the high esteem given to academics, the cachet of the institutions they represent, and the validation of their work by other academics is a truth that scholars with an interest in moving their work beyond the ivory tower cannot rest upon as a reason why we choose not to change how we are perceived by people like the reviewer of the proposal for this book that Michael mentioned earlier. In my conversations with many people I have met in academia, there is a delusion about the effects of their work on the communities that they research, or a perception that what that they do is enough to make a difference.

Michael: An important aspect of the misunderstanding between the two communities appears to arise from a lack of communication and perhaps from the lack of a venue where what now are two communities could meet to create a community of common purpose.

Eileen: Because education is such a complex process that involves a diverse set of stakeholders whose priorities differ and sometimes conflict, it's important to speak to different audiences through venues tailored for each. An attempt to write for all audiences by way of one source is likely to result in a mediocre product that does not serve any audience very well. Some sources do indeed involve scholars conversing among themselves. This venue is imperative in that it works to improve education by providing a base and sometimes a vision from which to move forward. Imagine the number of technological and scientific advancements we, the public, now enjoy because ideas were first debated and refined within the communities of science, technology, engineering, and mathematics before they were translated and transformed into products we now take for granted. The work presented in this text does both—provides the canvas for 21st-century science education and offers initial brushstrokes for what the masterpiece might be. The chapters, based in the real-life contexts of people, present detailed examples that address the query of "how-to" models that translate the abstract into the concrete in answering the question of "why," and clear arguments rooted in day-to-day living that respond to the "so what" indictment.

Chris: Once academics are honest about the fact that they are a part of an academic culture removed from practitioners and that they have worked to maintain this culture, it becomes possible to consciously create new practices that directly address the relatability or accessibility of our work. The first step in ensuring that teachers and students have, or seek, access to academic work is by ensuring that they are a part of all aspects of the research. If students and teachers are collectors, implementers and coauthors of research, the high esteem given to academics for doing this work comes to be shared with students and teachers. Furthermore, individuals within schools will begin to see that science education is a field where the research that is being produced is not separate from the classroom but rather is embedded in classroom practice.

Michael: In this way, academics, play an important role in the reflective explanation-seeking efforts that are the very condition for practical knowledge to evolve. As hermeneutic philosophers and ethnomethodologists alike know, explanation presupposes practical understanding; but practical understanding precedes, accompanies, and concludes explanation (e.g.,

Ricœur, 1991). True understanding on the part of the academics requires the same kind and level of practical understanding that makes teachers the successful practitioners that they are.

Chris: To subvert the focus on ourselves or our institutions as the primary focus of what we do, science educators must make efforts to change existent perceptions by transforming the nature of our practices. If teachers and students become co-teachers of college science education courses, are part of the research being produced, and are also producers of the research, discussing ways to connect them to the research after it has been produced would not be necessary. In fact, the creation of a genre within the field that focuses explicitly on practitioner-created research and the outcomes of their implementation of already produced research may serve as a way to begin this process. With a critical mass of scholars within the field that specifically focuses on the voice of the practitioner to re-create the culture of science education, the connecting of practitioners to the research will become a natural progression.

Michael: Katherine, like others, you are already drawing on your experiences in the culturing of knowledge to create curriculum that speaks to the needs of students in dealing with the science curriculum that—as SungWon, Miko and I point out in our chapter—is as foreign to students as are the societies and cultures of the countries to which we have immigrated.

Katherine: The border science curriculum I am creating has this understanding at its core. Since students, I have found, are already infusing their home- and community-based understandings into their science learning, and teacher practice, in fact, is already being influenced by those infusions, why not leverage those infusions and influences more formally so they can be recognized and talked about for what they are—a lived curriculum within an otherwise lifeless one? And let's be real: the lifelessness of the official science curriculum is not just a U.S. problem. From my research in science classrooms in Mexico I know that, due to the disembodied discourse that defines conventional science, this lifelessness can be pervasive. That is why developing my border science curriculum takes work with teachers on both sides. It's not that Mexican science teachers somehow naturally know how to do "culturing knowledges" better than U.S. science teachers. They don't. I find this the most powerful thing of all then—that the design of a border science curriculum can be transformative not just for the students and their families, but for teachers, both in Iowa and in Mexico who stand to discover different meanings of what science teaching and learning can be. So, returning to the central point of the reviewer's comments, in our work as academics we do surely need to mind and cross the theory–practice gap, but not with the illusion that, without us, such mindful crossings are not taking place. I believe students and teachers, perhaps without realizing it, make them all the time.

*

Michael: Some readers, like one of the reviewers, may agree with our underlying premise—that science and science education need to be rooted in the lived experience of the diverse human condition. But they may think that we replace past discourses with yet another jargon-laden hierarchy of "culturing knowledges." I wonder what we can say to those who would like to mobilize students' current and past experiences but who also know that scientific ways of knowing are incommensurable with everyday of culture and its characteristic forms knowing? How can we teach science all the while honoring, respecting, and capitalizing on the various non-scientific (indigenous, local, traditional) knowledges that are resilient and reproduce themselves precisely because they are useful resources for the people enacting them?

Katherine: In my earlier comments, I state that Mexican teachers don't necessarily know how to do "culturing knowledges" any better than U.S. science teachers. Generally, this point holds. However, the Mexican schooling system does create a particular site where "scientific" ways of knowing are put into direct and official juxtaposition with the "local" in a way that has great potential for helping us rethink how we do science education, particularly with Mexican newcomers. At the middle school level (*secundaria*), there exist technical schools (*técnicas*) that have developing in students a certain vocation-based knowledge and skill set as the primary goal.

Michael: This reminds me a lot of the secondary schools in the province of Quebec, where secondary students can take courses such as "Introduction to Technology" (where they learn handling common tools and operating certain machine tools, using and processing raw materials, and implementing manufacturing processes), "Home Economics" (where they learn basic techniques required in running a household), or "Manual and Technical Education I and II" (where the students learn how to use and work basic materials including wood, leather, cardboard, and plastic. The explicit purpose of these programs is the development of technical skills that students can put to use immediately or after graduation; and they can choose these courses even when they intend pursuing academic careers.

Katherine: The *secundaria* in the town of Angamacutiro that I talk about in my chapter has an agronomy (*agropecuario*) focus. There the curriculum revolves around giving students the knowledge and skills to succeed in farming and farming-related practices (or at least try to—in Mexico, small farmers face insurmountable economic challenges). What does this mean? It means that students' learning about reproductive systems in biology will be related to their knowledge that female and male cows are to be kept separate unless the motive is to have them mate, and that decisions about which cows to mate will be informed by students' studies of genetics that explain pure- or cross-bred status and predict height and girth. It means that the mathematics students learn will be used to calculate the amount of feed the pregnant cow needs to gestate a calf of a desirable size. It means that their studies in English will help them pronounce the names of the livestock breeds with which they are working (Hampshire, for example). The academic curriculum in this way is intimately linked to the technical curriculum in service of the cycle of work: grow crops to feed livestock, feed livestock to produce healthy livestock, slaughter healthy livestock to produce food (using all parts of the animal through food processing procedures), sell the produced food to buy seeds, sow the seeds to grow crops, and so on. And the academic-technical curriculum is rounded out by a civics curriculum.

Carol: I fully agree with this approach. Personally I've found that teaching science from the standpoint of pressing local problems is one way to teach science in the context of other knowledge systems. Interdisciplinary by nature, teachers and students can examine the issue from a variety of angles—none of which are accorded primacy, but collectively are essential. Too, this kind of curriculum requires a critical standpoint where students examine *how* material, sociocultural, and economic resources are leveraged in the search for solutions, and *who* is able to leverage these resources. Science in secondary and higher education suffers from being isolated from the other disciplines and students become myopic in their approach. For example, in one class that I taught in New Mexico we examined type two diabetes, a medical epidemic in Southwestern communities. Students and I studied the science of cell metabolism, but also shared our personal knowledge of the disease from our own experiences, and had traditional healers also speak to us. We also situated our discussions within a social and political history of commodity foods, and the loss of agrarian traditions. Much like Katherine and Hannah's "Border Science Pedagogy," science was but one of many aspects of the problem we studied.

Katherine: What is more, the students learn responsibility and respect because, in order for the cycle to work, there must be cooperation among them. This is a curriculum that exists precisely because it *is* a useful resource for the students who will engage in very real ways the academic/technical/civic practices it contains. And it *does* bring students' everyday forms of knowing into commensurability with the "scientific." What I think is important is to put these Angamacutiro teachers into conversation with other teachers, both in the US and Mexico, so we can understand more about the balance they've struck between the work of "culturing knowledges" and "science." It seems to be working. Out of a graduating class of 80 this year, 65 are going on to high school (*preparatoria*). Given that high school is not obligatory in Mexico and that we are talking about students from a poor, rural context, this rate is pretty remarkable. While talking about a "vocation-based" curriculum, particularly in relationship to non-dominant communities, is problematic and perhaps even taboo in the US, I think the Angamacutiro example demands that we attend to the ways in which practices that make connections to home- and community-based experiences can serve students and their families and communities well in their science learning and living. Of course, with the continued destabilizing effect of NAFTA on the Mexican small farmer, the vision of self-sustainability offered by the Angamacutiro school becomes less feasible. This deserves to be addressed in the curriculum as well, and can be, through a cultural studies approach to science education that situates science practices and their relationship to societal infrastructure (politics, economics, technology), in a sociohistorical perspective.

Eileen: These kinds of experiences exist in the US and are often included in the vocational "tracks" of the school curriculum. I see the benefits of these tracks but it is also important to consider the inequities and unequal opportunities that result from them—one reason why vocation-based curricula with respect to non-dominant groups are problematic and taboo in the US. Perhaps these tracks do provide a mechanism through which canonical scientific knowledge connects with local knowledges, but these experiences often lead to life positions with less power, less authority, and fewer material and social privileges. Without a more critical component, such curricula, although connecting privileged to local knowledges, serve to limit access for non-dominant groups, keep non-dominant groups in their subordinate positions in society, and reify the status quo.

SungWon: Katherine's comment emphasizes the point of this book, which is that cultural translation/hybridization appears at every practical instance of teaching and learning science and this translation work leads to deeper understanding of cultures and identities that different ways of talking allow. From my experience of conducting the research project, I learned that respecting different ways students experience the world requires the articulation of new cultural possibilities that educational practice makes available in the fullness of life. In this sense, educational practice is to participate in the active translation of culture and knowing.

Michael: The term translation often implies two entities (usually language, sometimes mathematical representations) that are foreign to one another. Some readers may therefore be strengthened in their distinction between two forms of knowing, which, from my perspective, cannot ever be pure but are hybrids in themselves.

Eileen: Inherent in your original question is the premise that scientific ways of knowing and other ways of knowing are exclusive, and that the scientific way of knowing is superior. Unfortunately, the premise is dominant and to alter this premise should be a core focus for action.

Chris: There is a prevalent misconception when it comes to discussions about scientific ways of knowing and their incommensurability with the everyday culture of certain populations. The fact that there are differences between the culture of some students and that of science does not mean that there are no points of alignment between them. In fact, there are points of connection between all cultures that have the potential to create new understandings of each of the cultures that come together. If we continue to perpetuate the ideology that there are these insurmountable differences between certain cultures and science, we defeat the purpose of trying to have students from these cultures become successful in science. The work of the science educator in an effort to "mobilize students' current and past experiences" within science education is to investigate the aspects of student culture that align to science and science teaching and vice versa. A second responsibility of the researcher is to facilitate the creation of scenarios where educators can begin to find these points of connection between the culture of their students and that of science for themselves.

Eileen: And the success of such action is less dependent upon the methods and more upon the attitude in which it is implemented. When I think about the different ways of knowing and respecting them, the parts of the human body and how they function independently and in concert come to mind. Because my eyes differ from my ears in their nature and in their functions, would I dare say that one is better than the other thereby restricting or eliminating the use of the one I designated inferior? I prefer to maximize my capacities and independently or in concert utilize my eyes and ears for their distinctive functions. I apply the same logic with ways of knowing. Scientific ways of knowing and other ways of knowing need not be exclusionary, or one deemed superior to another, but they can be parallel in the ways in which they are salient in people's lives. This parallelism does not mean that contradictions and conflicts do not exist, but these contradictions and sites of incommensurability can be highlighted without devaluing—it is this posture that is necessary and crucial in teaching the ways in which we as humans come to know. Canonical science is one of those ways.

Michael: Your point about knowledge being part of a hybrid system, which allows these forms of knowledge not only to be different but also to be contradictory, has already been made in the early 19th century by Friedrich Wilhelm Joseph von Schelling (1994–2007) in an essay entitled "On the nature of philosophy as natural science." He uses the same analogy of the human body. He suggests that the contradiction between different systems of knowledge is the *condition*, the material, for the possibility of knowledge as a system.

Chris: Let me continue with the philosophical slant you are taking. The chief issue that we should bring to readers is a focus on the embeddedness of school science to a specific cultural ideology that necessarily separates and advantages itself. It is necessary for us to make clear that while it may be obvious that science education needs to be rooted in the lived experience of students in schools, the realization that the closeness of the culture of students to the Cartesian/Newtonian/Baconian ideology that school science is comprised of is a strong determinant of their success in the classroom. Both students whose ways of knowing are closer to this more Western ideology and those from more diverse cultural backgrounds will benefit from a focus on their specific cultures in the classroom. However, the point to drive home is that those whose cultures align closely to that of school science have an unfair advantage over their counterparts because their knowledges are already embedded in the cultural contexts of the classroom. The issue is not whether or not student culture should or should not be brought into classroom. The answer to that question is obvious. What needs to be focused on is a leveling of the playing field where knowledges that are not traditionally represented in the science classroom are given an opportunity to be expressed within it. The work that has been done in this book is a first step to accomplishing this goal.

*

Michael: An issue related to that of the cultural appropriateness of science education is that of place-based science education, which has the intention to draw on students' familiarity with their surrounding world as a starting point for learning. Here, too, there will be an inner contradiction because science has been formulated in a way that it works everywhere the same independent of place. But interestingly, it does not do so on its own, as Bruno Latour (1988) showed in his *The Pasteurization of France*. It is not science that is independent of place but that places have to be configured for science to exhibit itself in a particular way. It was only when Pasteur changed the stables to resemble the conditions of his laboratory that his science actually came to work. Ms. Herlton, Anselmo Quam, and the workers in the meatpacking plants all have place-based knowledges and experiences. Is it possible to mobilize such place-based knowledges in the service of science education? Or do we have to change the very nature of science?

Katherine: Revisiting my previous responses, I think I'm clearly critical of how "science" has been constructed to begin with. Instead of relegating place-based knowledges as peripheral, second-class epistemological citizens, then, what I want to be part of is a movement to exert pressure on the center. I'm not naïve, though, about what the outcome is likely to be, nor the contradictions (perhaps hypocrisy) of my taking up that position. "Science" is a huge industrial complex, one from which I, the other contributors in this book, and many, many others benefit each and every day. There is no question that endeavors guided by "science" have yielded enormous social change, much of it beneficial in ways large and small. Professionally, I invested most of my young adulthood in assimilating myself into a scholarly community's norms for thinking, reading, writing, listening, and speaking—norms that were guided by, even if they were also dutifully critical of, (social) "science." And the work that I accomplish through this socialization, I hope, is helpful to my students, other scholars, and the communities I research. And, personally, as the child of a mother suffering from Alzheimer's and the sister of two schizophrenics, I look to "science" to spare myself from the former and my children from the latter. So I know to tread lightly when it comes to this question. But I am convinced that Sandra Harding (1998) had the answer when, in "*Is science multicultural?*," she asserts that standpoint epistemologies, a concept akin to "culturing knowledges" in that it theorizes from the experience of the marginalized, actually enhance "objectivity."

Michael: Which is the point the already-mentioned Schelling makes. The contradiction in forms of knowledge has an objective reason, and building an epistemological system on and including these contradictions leads to a more objective science of philosophy. This means that objective knowledge is not one perspective but in fact is multiperspectival.

Katherine: Applying multiple perspectives to an issue or question is the only way we can ever actually "know" it. "Science" knowing is incomplete if it doesn't recognize other forms; it isn't, to use a qualitative research term, "trustworthy." It may not even be, to use another, truly "catalytic"; that is, it may not actually impact in productive ways the lives of the people with whom it purports to be most concerned. I have become intensely interested in the idea that biodiversity's strength in natural settings may have a parallel in the social. Corporate America seems ready to pick up the message that bringing together people with different perspectives actually leads to innovative problem-solving. Why aren't U.S. schools and science teachers and scholars similarly persuaded? The very nature of science would indeed need to change to encourage their participation in the idea of a multidimensional science framework; without such change, "other" ways of thinking through science will always be just that.

Eileen: I do not think we need to change the very nature of science; I think more needs to be done to dispel the myths about the nature of science and to highlight and critique its nature, which Chris did nicely in his earlier comments. Science, like all ways of knowing, is a placed-based human construction. It is performed by humans who are situated and influenced by place, and science exists and is used for specific purposes. To mobilize placed-based knowledges and experiences, it is crucial to explicitly highlight and possibly expand the purposes of science education and to overtly position science teaching within them. Such an expansion would include the purpose of person formation discussed by Katherine and Hannah in their chapter.

Chris: With the injection of place-based understandings into science education, opportunities are provided for science education to incorporate the understandings of populations that are traditionally silenced within schools. Furthermore, science education is allowed to reach a place where the science being taught is more indicative of true science.

Carol: I, too, strongly advocate that science education be developed from a foundation of place-consciousness. Place is not just a physical setting, but also embodies the political history, as well as the social and cultural context. Christopher Emdin's approach to using hip hop in the science classroom exemplifies one way that place-consciousness can open up new avenues for students to participate in science. The intersection of science and hip hop captured the struggles of Black youth in the social, political, and economic urban context. It's not so much changing the nature of science as it is opening up science to the concept of standpoint as being integral to sites of knowledge production.

Chris: The variation between science and school science becomes a particularly important issue here. Whereas school science reflects a discipline that is fact-laden and based on the existence of certain facts and principles, science has evolved to include a consideration for the nature and epistemology of the discipline in a more holistic and complex way of looking at the world. The more contemporary view of the nature of science makes allowances for the mobilizing of place-based knowledges into science, science education or more specifically; yet the self-nominated gatekeeper institutions that separate the layperson from science do not make a space for knowledges of populations that are viewed as not able to succeed in science to be expressed.

SungWon: But people are able, as expressed in their knowledgeabilities. I agree with the demands for praxis-centered approaches to the rigorousness of science and generalization. I remember the suggestion that "the difference between ideology and science is the difference between treating those concepts as the primitives of theory and treating them as sites for exploring the social relations that are expressed in them" (Smith, 2004, p. 446). I think my chapter exemplifies that embodied translation of the local constitutes the central aspect of learning science, and that this translation involves the amplification of the particularities of the local rather than the deletion of them. Perhaps the inner contradiction related to the issue of generalization would be articulated not between different forms of local knowing but between a specific form of knowing and the translation that it cannot but undertake at the cultural encounter.

*

Michael: One of the assumptions made by a culturally sensitive science education is that there is a possibility to preserve and sustain the values and epistemic qualities of the everyday and scientific cultures. The concept of the mêlée contests such a conception because it explicitly articulates the increasing hybridization that arises in cultural encounters. Even the concept of the "third space" suggests that what students enact is not science but some hybrid culture that is neither home nor scientific culture. If this is the case, what does this mean for anybody's aspiration to teach "authentic" science, that is, science and scientific knowledge characteristic of laboratories rather than of the everyday world where it is inherently modified for the purposes at hand?

SungWon: I agree with the significance of the concept of the mêlée . It leads me to think about the "authenticity" of teaching and learning science neither in relation to one local culture/language nor to "anything goes," both of which cannot help but delete the gap between the curriculum (as prescriptions) and the lived experience or between the cultural categories that the gap has presupposed. Rather, I think that "authenticity" arises from the translation that the practice makes, and the associated, new cultural possibilities generated in the course of it. In my chapter, I attempted to show that the passivity and the heterogeneity of our Self lie at the heart of this authenticity.

Chris: For a person who wants to teach "authentic" science, it is important to recognize that what is named as authentic science should be viewed as a brand of the multiple ways of knowing that encompass science. This type of science should become just another brand of science (like hybridized versions) to be taught, learned, and understood.

Eileen: Writing about authentic science in my own chapter, I adhere to premises of culturally sensitive science education and question the idea of hybridization, the extent to which it occurs and in what domains. Oftentimes, hybridization is presented as though it automatically and always occurs and to the fullest extent. I do not believe that every cultural encounter results in hybridization; structures like those described in Katherine and Hannah's chapter, and experiences such as those described by Chris and me in our chapters prevent it and individuals, to varying degrees, can thwart it.

Katherine: As a postmodern, poststructuralist thinker, the boundaries between "authentic" and "everyday" science are inherently problematic. I absolutely resonate with the idea of the mêlée because it accounts for what I have observed in the science instruction of Mexican newcomers where they, with their (White) teacher, come to participate in hybrid science instruction practices that bridge their cultural and linguistic backgrounds. In this respect, I've come to think of the notion of "authentic" science as, to invoke Judith Butler (1990), a "regulatory fiction"; it doesn't really exist, but striving towards it helps achieve some identified, supposedly desirable end.

Michael: You are kind. I might have extended the reference to regulatory fiction and added, "to be subverted" to yield the statement, "'authentic' science is a regulatory fiction to be subverted."

Chris: I take issue with the notion that the results of the coming together of science and the culture of a certain population (whether through hybridization or within a third space) results in something that cannot be named as science. My issue is not because I do not believe that what is created is not different from any of the previous cultures previous to their coming together. It is with the fact that the naming of a more complex understanding that is other than science as not a brand of science or scientific often equates to a perception that it is less than science or is an adulterated form of science that has less value than the original. The reality is that we have inherited a view of science as the ultimate way of knowing that automatically relegates anything other than science to a lower position than it should have. . .

Michael: . . . where I would want to see the "we" as one part of society engaged in a politics of domination, philosophers such as Schelling have argued for a long time that hybridity of knowledge is a precondition for anything like science to emerge and exist. Without this hybridity and inner contradiction of mutually exclusive forms of knowledge, there is no knowledge as system.

Eileen: So, from the position that challenges hybridization, achieving authenticity that retains the integrity of different ways of knowing is to acknowledge and accommodate the multifaceted, dynamic, and context-influenced nature of what it means to be authentic. In essence, authentic science, to pick up the concept that SungWon and Chris already used, is largely determined by the purposes of science. What is typically promulgated as being reflective of the scientific enterprise is authentic science if the purpose is to prepare someone to function within scientific communities that currently exist. This authenticity related to the purpose of preparing someone for membership in present-day scientific communities in no way lessens the purpose-driven authenticity in the lives of Ms. Herlton, Anselmo Quam, and the workers in the meatpacking plants. As with the parts of the human body, the nature and function of authentic science is not the same from one context or place to another and each in their complexity and distinctiveness is needed for optimizing our understandings of the human experience.

Carol: For me, then, each chapter in our book portrays knowledge production and emphasizes the local-ness of practice, the "space" in third space, if you like. Your use of mêlée , Michael, is somewhat akin to what Turnbull (2004) describes as a "motley," a messy assemblage, or knowledge space consisting of sites, people, and activities. Turnbull argues that Eurocentric science has acquired a sense of coherence through the work of making equivalences and connections. But as this knowledge space has been elevated in our culture, it has taken on an unchallengeable naturalness, and becomes seen as "common sense," where the labor of standardization and homogenization becomes an essential, but an unacknowledged aspect of its maintenance. In each of our stories, I see how a historical account has helped each one of us understand the ways that Eurocentric science and these other knowledge spaces developed, how they bump up against one another, *and* how they can become sites of resistance. For me, authenticity comes with helping students and teachers to see how this motley of practice and knowledge production functions in their lives, in the laboratory, and across other contexts.

Katherine: Let me return to my earlier comment. What I think this volume is calling into question is precisely what I denoted as the "desirable" part. Each of the authors here envision a decompartmentalized ontological and epistemological approach to teaching, learning, and writing science that rejects claims to its "authenticity." But I think the idea of "applied" or "everyday" science, in a romanticized way, also merits interrogation. I am reminded of Michael Pollan's (2006) discussion, in *The omnivore's dilemma*, of the discovery of the technology of nitrogen-splitting. The result of this innovation, which no doubt occurred through "authentic" scientific work in a laboratory, was two notable applications: the fertilization of croplands to increase yield and the gassing of Jews in concentration camps. The first application was apparently life-giving (the ecological destructiveness of the practice has since come to light) and the other life-extinguishing, yet they flowed from the same scientific source. So, clearly, application, not just the "science" behind it, is an important part of the issue. Whereas the horrors of the concentration camps are behind us, ecological destruction remains. And this quiet destruction may be the most difficult to remedy precisely because the agents of destruction are friendly farmers and the beneficiaries cost-conscious consumers who want fruits and vegetables despite the season. Taking into account this tripartite of the technical, applied, and consumptive sides of science pushes "culturally sensitive" science education from just "politically correct" to socially responsible. That is where I would like to see it go, not as an add-on feature of the curriculum, but as a core principle. If we are to have the democracy envisioned by social dreamers of this era and others, we must get down to the business of preparing students to be critical thinkers about the decisions that, once taken, become the starting point for the next generation. Science education has an important role to play in conjuring the future.

References

Butler, J. (1990). *Gender trouble: Feminism and the subversion of identity.* New York: Routledge.

Harding, S. (1998). *Is science multicultural? Postcolonialisms, feminisms, and epistemologies.* Bloomington: Indiana University Press.

Latour, B. (1988). *The Pasteurization of France.* Cambridge, MA: Harvard University Press.

Pollan, M. (2006). *The omnivore's dilemma: A natural history of four meals.* New York: Penguin.

Ricœur, P. (1991). *From text to action: Essays in hermeneutics, II.* Evanston, IL: Northwestern University Press.

Schelling, F. W. J. von (1994–2007). Über die Natur der Philosophie als Naturwissenschaft [On the nature of philosophy as natural science]. abc.de Internet-Dienste. Retrieved July 22, 2008, from http://gutenberg.spiegel.de/schellin/essays/natur.xml

Smith, D. E. (2004). Ideology, science and social relations: A reinterpretation of Marx's epistemology. *European Journal of Social Theory, 7*, 445–462.

Turnbull, D. (2004). *Masons, tricksters, and cartographers: Comparative studies in the sociology of scientific and indigenous knowledge.* New York: Routledge.

Part II

Othering the Self, Selfing the Other

Introduction to Part II

> Knowing [savoir] (but based on which knowledge [savoir]?) that from now on the subject of knowledge can only be some*one*, and like every*one*, someone of *mixed blood.* (Nancy, 1993, p. 10)

Two almost intractable problems of educational research have been related to the questions of (a) how to make research, often abstract and referring to nobody in particular, relevant to the real lives of people and (b) how to generalize from the experience of individuals to the experience of others. These problems, it appears to me, are the outcomes of a particular way of looking at and seeing the world generally, and the relationship between the general and particular specifically (Roth, 2009).

Self, selfing, to self: terms used in the biology of reproduction, where the pollen of a flower is placed on the stigma of the same flower or another flower of the same plant to produce seeds and seedlings that exhibit desirable characteristics. Selfing the Other then means to reproduce this Other through a sort of self-fertilization. This necessarily happens as the Self cannot be other than the perspective from the Other on the reflecting consciousness.

Stories of experience necessarily are stories in a language, which never is the singular language of a person, but always and already the language of the other, from the other, and for the other. The language comes to us in particular narrative forms, genres; and these, too, come to us from the other and are produced for the other. To be intelligible at all, the narratives themselves have to have all the resources that allow another person to make sense of them, so that both the form and content of any narrative designed for the other are intelligible to this Other and therefore merely realizing a cultural, general possibility.

In speaking, the Self others itself, reproduces a Self in the image of the Other, and, thereby Selfing (self-reproducing) the Other. Self-consciousness is a form of consciousness, which, as its etymology suggests always and already is *consciere*, a knowing (Lat. *scīre*) that is common (Lat. *con-*).

In this section, I have assembled three chapters that are fundamentally grounded in the experiences of the authors: Angela Calabrese Barton (chapter 8) writes about the experience of a nagging doubt about the health of her second child, who, as it turns out, has *torticollis*, a twisted neck; Karen Tonso writes about the culture of an engineering school from the perspective shaped by her own experience of going through such a school and having been an engineer for 19 years (chapter 10); and my own story (chapter 9) is about my battle with a chronic illness that the physician ultimately named "chronic fatigue syndrome slash fibromyalgia." In each case, we can learn from the chapter more than the particular aspects of one concrete lived experience. In each case, we can see general cultural possibilities realized in a concrete manner, possibilities that also exist for others. In each case, we see people struggling, with a medical symptom, an illness, or with the processes of becoming as a new member in a culture. In each case, neither symptom, nor illness, nor growing pains are singular as phenomena, because others have had experiences with family resemblance before and others will have similar experiences after.

By focusing on the cultural dimensions concretely realized in each of the stories, we learn about the culture as such. We learn through the Selves of the individual (authors) involved about Others, who in their ways, do and can realize the same and similar possibilities.

Othering the Self and Selfing the Other therefore are but two sides of the same coin, aspects of the very possibility of living, being conscious, and having something like a language to communicate.

All three stories also provide us with at least glimpses at "the interrelation between thought and language and other aspects of mind" (Vygotsky, 1986, p. 10), as they allow us to see that thought processes, experiences, and intellect are not independent of affect, the separation of which was a major weakness in Vygotsky's day and has continued in present-day science education. When intellect is segregated from affect, "it makes the thought process appear as an autonomous flow of 'thoughts thinking themselves,' segregated from the fullness of life, from the personal needs and interests, the inclinations and impulses of the thinker" (1986, p. 10). Thought then becomes a mere, "meaningless epiphenomenon incapable of changing anything in the life or conduct of a person or else as some kind of primeval force exerting an influence on personal life in an inexplicable, mysterious way" (1986, p. 10). It becomes impossible to understand the causation and origin of our thought. Yet, as the chapters show, what we do and how we do it has its very origin in the affective dimensions of life, the concerns for others, experiences of being ill, and having experienced with one's own body what it means to be and become an expert (engineer).

References

Nancy, J.-L. (1993). L'éloge de la mêlée (Eulogy of the mêlée). *Transeuropéennes, 1*, 8–18.

Roth, W.-M. (2009). Phenomenological and dialectical perspectives on the relation between the general and the particular. In K. Ercikan, & W.-M. Roth (Eds.), *Generalization in educational research* (pp. 235–260). New York: Routledge.

Vygotsky, L. S. (1986). *Thought and language.* Cambridge, MA: MIT Press.

8 Mothering and Science Literacy

Challenging Truth-Making and Authority through Counterstory

Angela Calabrese Barton

When I became pregnant with my first child, I was told by many people, "You need to read *What to expect when you are expecting*." So, like the dutiful new mother-to-be I went out and purchased the book, and about five others describing the ins and outs of pregnancy and babies. After all, I am a science educator and knowledge of how the body works is interesting to me, especially when it is my own body! But I quickly became frustrated because the tone of most of these books felt not only paternalistic, but also essentialized, as if all pregnant bodies worked the exact same way. I should feel some morning sickness, and I should sleep with crackers next to my bed to eat when I woke up to alleviate any nausea. I should not exercise rigorously, and I should stay away from non-healthy foods like ice cream. Now, I understand the "science" behind these recommendations, but these generalized claims did not fit my world. The thought of crackers made my stomach churn, and running an easy 4 miles made me feel well. So, I expanded my search for information, including joining an on-line community of other mothers. A subset of us then later formed our own private forum close to five years ago because, as we got to know each other better, we wanted a safer space to share our stories and experiences about our children and ourselves. Since that time many of us have met "in real life" though our friendships survive on-line.

Whereas this community has served as multifaceted site in terms of the reasons why the women post there, as a mother who is interested in science learning and science literacy, I am often struck at how this space has become a knowledge-generating community that lacks both the essentialist and paternalistic overtones of those books I purchased five years ago. The knowledge of this community, which is distributed, personal, contextual, and often contested, has not replaced the world of doctors or formal medicine, but has become one of the filters I use to understand myself and my children's health. In this chapter I explore how this on-line community of mothers has provided me with a collective wisdom that is personal, contextual, and contested. To present this story I first turn to standpoint theory to make a case about how science literacy can be understood as a process of weighing, comparing, agonizing over, and sifting through competing experiences and knowledges. Then, I tell a story of coping with and in science-mothering—*torticollis*—and the role that this on-line community played in coping.

Science Literacy for Coping

Science literacy for all has been a dominant theme of the US-based science education reform initiatives since the mid-1980s. At the time the phrase "science literacy" gained momentum, it referred to the "big ideas" that shape our understanding of the natural and human made worlds as well as "knowing that science, mathematics, and technology are human enterprises, knowing what that implies about their strengths and limitations; and being able to use scientific knowledge and ways of thinking for personal and social purposes" (AAAS, 1990, pp. xvii–xviii). In other words, science literacy includes knowledge of science and the nature of science as well as knowledge of science as a social enterprise. Yet since its introduction there has been wide debate around what really constitutes science literacy, for whom and under what conditions (Brown, Reveles, & Kelly, 2005). The extent of the debates have included what big ideas matter and why, how understanding might be fostered and demonstrated, and how students might turn thinking into a practice of science. The American Association for the Advancement of Science (AAAS, 1993) published "benchmarks" for science literacy, which reflected "what" students should know with respect to the big ideas in science. The National Research Council (1996) published a set of national science education standards that captured the content and skills that school students ought to acquire. More recent research in science education has focused on "learning progressions" to capture the nuances in how we frame what it is that students ought to know, how, and when. Cutting across all of these debates is a presupposition that such understandings matter in how individuals learn to engage in democratic society. However, little attention has been given to the structure or forms of science knowing that might support this lofty goal.

Following from this point is another side to the science literacy debate. This other side, which has gotten less attention here in the US, but has been more at the center of the science education storms in the UK and Canada, is the extent to which science literacy should also explicitly incorporate notions of citizenship and democratic participation. For example, while schools should be concerned with fostering understandings of basic foundational ideas (similar to AAAS's big ideas) and understandings of the nature of scientific inquiry, they should also foster a regular practice of consuming science information through reading about and evaluating scientific ideas in the public domain. Whereas this way of framing science literacy does not drastically differ from what AAAS put forth ten years before, it does re-position the value of individuals gaining a propensity for consuming science in ways that are critical and that matter to them.

Other science educators have been more far-reaching in their view on civic science literacy or science literacy for citizenship by arguing that what is important is not just "what" individuals know or how they come to know it, but how such knowledge is brought to bear in the decision-making process and for what purposes. For example, real literacy in science involves deep understandings of the human and social aspects of the production of scientific knowledge, and a critical assessment of knowledge and sources of knowledge, so that individuals can be thoughtful decision-makers on socio-scientific issues that dominate life in the 21st century, such as gene therapy and cloning. Others push us even further to consider how the process of coming to make decisions on larger socio-scientific issues involve "collective praxis" and "situated knowing/doing," taking science literacy out of the individual realm and into a public practice (Roth & Lee, 2004). Nonetheless, cutting across these arguments for science literacy for citizenship is the concern that reform documents like that of *Project 2061* do not go "far enough" in delineating "how" it is individuals might take what they know in order to use it towards participation in democratic society.

While much of the work around science literacy for citizenship has focused on "how" scientific knowledge and habits of mind might be brought to bear on public decision-making, what has garnered almost no attention is science literacy for coping with life. In this book, Roth raises the question of how science and science education might assist everyday folk in and with the problematic situations that they face in their lives. Here, science literacy involves not just what people need to know but how that knowledge is fashioned by location.

It is interesting to note how this more personal, highly contextual way of thinking about the role and value of science literacy has been absent from nearly all debate in the science education community. I submit that this is probably because, in global politics, individual well-being is not perceived as carrying the same economic weight as national security and economic vitality. I also submit that this is because to frame science literacy as about coping with life at once makes science accessible to all individuals and positions the situated nature of knowing as a critical (if not the most important) dimension to the nature of knowing and scientific knowledge. This stands in stark contrast to the positivistic history upon which the growth and concentration of power in the sciences has been built. I draw upon standpoint theory to delve into this point in more detail.

Standpoint theory has deep roots in feminist and Marxist thought, with focused attention paid to how understandings and explanations of the world are always situated and written from the "standpoint" of the social agent. As Harding (2004a) explains:

> The women's movement needed knowledge that was for women. Women had long been the object of others' knowledge projects. Yet the research disciplines and public policy that depended upon them permitted no conceptual frameworks in which women as a group became the subjects or authors of knowledge; the implied "speakers" of scientific sentences were never women! (p. 29)

Yet, standpoint theory gained footing in feminist worlds because it offered more than simply explanatory power. In addition to being a theory about epistemology, it is also "a philosophy of natural and social science," a "methodology," and a "political strategy" for empowering oppressed groups—fields and projects which "conventionally are supposed to be kept separate" (Harding, 2004b, p. 2).

Yet standpoint theory has been dogged with controversy from its inception. As Wylie explains:

> Standpoint theory may rank as one of the most contentious theories to have been proposed and debated in the twenty-five-to-thirty-year history of second-wave feminist thinking about knowledge and science. Its advocates as much as its critics disagree vehemently about its parentage, its status as a theory, and crucially, its relevance to current thinking about knowledge.
>
> (2003, p. 27)

Standpoint theory is controversial because it espouses that politics are an important element of knowledge production, and that the concerns of oppressed groups are not just social and political in nature but intimately linked to knowledge production itself and its relationship to power. However, Harding (2004b) argues that the controversialness of standpoint projects is what has given them value and power: "Standpoint claims about the origins, nature, and role of group consciousness in the production of knowledge should be controversial" (p. x).

To return to science literacy for coping with life, what standpoint theory helps us to do is to uncover the subjectivity of truth-making and to re-position the meaning of epistemic privilege within and across communities. In particular, standpoint theory offers us conceptual and personal spaces for hearing and telling "counterstories"[1] in and about scientific knowledge and knowledge production, allowing us to imagine a world beyond normative claims. Science literacy can be understood as a process of weighing, comparing, agonizing over, and sifting through competing experiences and knowledges.

In other words, standpoint theory allows us to turn current inceptions of science literacy on its head, for science literacy must move beyond what one must "be able to know and do" to account for how such knowing and doing intertwine epistemological, methodological and political activity. It shows us how becoming scientifically literate may be about developing understandings and habits of mind and use those for democratic participation, but it is also centrally about the personal process of coming to know in highly contested world.

Science Literacy and Mothering

I have always fashioned part of my identity as that of a scientist. I resonate with wanting to question and understand the world in systematic and rigorous ways. I take some pride in having some confidence in being able to ask questions about things that do not seem to make much sense. So why is it that having to pursue the health-related of issues of our children can be so damn difficult for those of us who identify with science? Why is it that when I persist in asking and re-asking a question that seems to make perfect sense to me, I feel patronized by the medical establishment? Why is it when I try to contextualize my own or my children's experiences, I tend to get generalized responses that seem to pertain to everyone else but us?

Five years ago I joined a closed on-line community of about 40 women that was created for and run by a group of women in their 20s and 30s who were new to motherhood. Most of us joined because we had just had or were about to have our first child, whereas some participants were hoping to become pregnant or adopt or already had multiple children. Many of the women are "stay-at-home" moms, though many of us work full or part time. Most of us are from the US though some are not. Many of us have college degrees, although some do not. After five years most of us know at least some members in real life while some maintain only e-relationships.

Whereas conversations among us range from the mundane—what's for dinner?—to the deeply serious—one mother's struggle with a toddler with cancer—the spirit of the group revolves around building a community knowledge base that is distributed, personal, contextual, and often contested. In this section I explore how this on-line community of mothers has provided me with a collective wisdom that is personal, contextual, and contested by chronicling a story of coming to know around my children's health and well-being.

Frankie's Torticollis

My second daughter was born in June 2005, a healthy girl weighing in at 8 pounds. My pregnancy, while trouble free and relatively easy, was different from my first. My hips, and in particular my left hip, ached horribly. My husband and I joked how little Frankie must be perched on that side of my pelvis.

We also joked about how Frankie came into the world ready to run. She kicked incessantly in the womb, and she seemed to be most happy when I was up and about. Seeing as how we lived in a large urban center, we even walked to the hospital when I was in labor, and she was born 45 minutes later upon our arrival. I'm not a very big person myself, and my doctors and nurses agreed that she felt like she was going to be "6 pounds or so." So when she literally came flying out of me at 8 pounds we were all somewhat astonished. Even our doctor was surprised and remarked that Frankie was a master magician managing to hide away 2 pounds.

Now little babies, according to the books we are told to read, are not supposed to be able to lift their little heads until they are about four to six weeks old. As parents, it is our responsibility to be diligent in providing proper support for the neck and head or else it will easily bob to one side or the other, potentially causing damage to the fragile spinal column and bones.

Frankie was not our first baby, so none of this was new to us. Yet, whenever we set Frankie down in her bouncy seat or even put her on our bed or the floor, her head would also tilt to the right. If we put extra supports on that side to keep her head in place she would finagle her whole body into a new position so that her head could tilt to the right (see Figure 8.1).

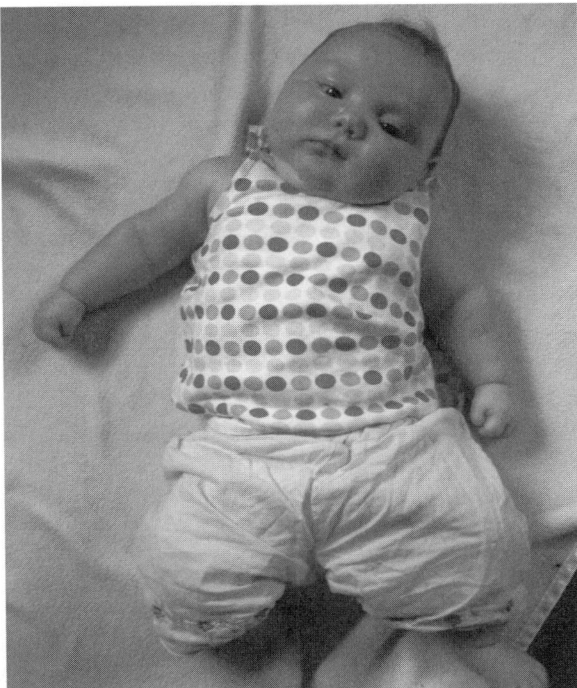

Figure 8.1
Frankie, 4 weeks old.

At her six-week "well baby" check up, my pediatrician did not think much of it because at the doctor's office my daughter never quite "performed" in a way that would demonstrate this tilt. Furthermore, she told us that a baby's head always tilts or bobs in many directions because the neck is simply not strong enough. Whereas our pediatrician seemed interested she also seemed unalarmed. After all, babies at six weeks are really too young to hold their heads up. We need to be more diligent about supporting her head. But of course! Anyhow, my husband and I thought that Frankie already possessed a great deal of neck strength for her six weeks of age!

But, the tilting continued, even as her neck muscles got stronger. And, it nagged my husband and me. So, a few weeks later, when she was about eight weeks old, I took a few pictures and showed them to my on-line friends (see Figure 8.2).

In my first post, I wrote:

> So, I have been a bit worried about Frankie. She always seems to tilt her head to one side. Even when we prop it up, she finds ways to maneuver her body to keep her head at a tilt. Maybe I am just imagining this, but dH noticed it too. In fact he noticed it first and asked me what I thought. Here is a picture of her. Do you think I should be worried?

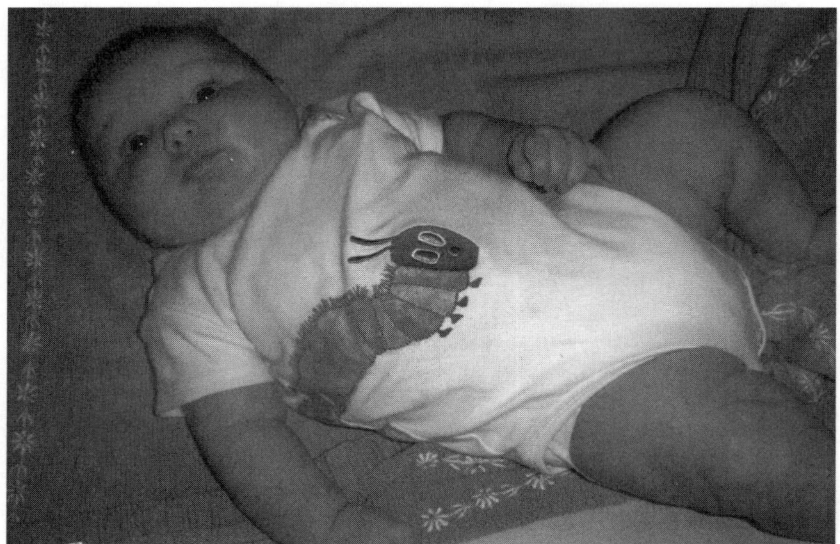

Figure 8.2 Frankie, 8 weeks old.

Some of the moms wrote back and said they had no idea but maybe I ought to check with my doctor if I was worried. Hmmm. I had already done that. What else? Another mom wrote that her daughter's head seemed to always tilt to one side too, and her pediatrician agreed giving the mom gentle stretches to do on the baby. Another mom said that she had heard of other babies with a similar problem and that they needed physical therapy. Four other moms just said {{big hugs}} and similar sentiments.

Two moms, however, implored me to get over to the pediatrician's immediately and demand a request to see a pediatric neurologist. Hmmm. Just that phrase—pediatric neurologist—made me nervous! After conversing with these two moms in more depth, it turns out that both of these women who had responded to me had a child who had had torticollis, but neither child was diagnosed early enough to prevent some head bone misalignment or plagiocephaly. One went undiagnosed until 12 months, and the other till 9 months. While torticollis is, in most cases, not life threatening, it can affect how the head grows. Both mothers ended up having to have their children wear plastic helmets to reshape the head, to correct for the damage done by the torticollis.

One of the moms told me that her child's torticollis (and plagiocephaly) went undiagnosed for 12 months before the pediatrician first noticed the plagiocephaly. She told me that she noticed her child's head tilted, but did not worry much about when the pediatrician did not seemed concerned. The other mother told me that her child's torticollis was noted but not of concern until six months of age when her child began to exhibit signs of plagiocephaly, something his doctor did not expect so early. I began to wonder. Were these mom's experiences typical or atypical? Should I not worry about Frankie and wait until my pediatrician also registers this as a concern?

Torticollis. I had never heard of it before, and my gut reaction was, well, wouldn't the pediatrician have been concerned about it when we visited two weeks prior if it were a real issue?

What ended up unfolding was an awakening on my part along two lines: First, that mothers, collectively, have a knowledge of children's health that is nuanced and deep, and that often flags concerns before the medical profession might clinically notice them. Second, that clinical models for assessing children's well-being are simply that: models. They are not perfect and should not be viewed as an exact description for all children's development.

So I began to engage in some drawn-out Internet research on torticollis. Actually, first I had to learn how to correctly spell the word. What I learned from the Internet surprised me. Not only did I learn that this condition was not that uncommon—0.5% of all children are born with some degree of torticollis—but also I learned that, with early intervention, more invasive physical challenges, such as the development of plagiocephaly, could be prevented:

- Torticollis is a condition which affects the neck and spine in 1 in 200 infants, and two thirds have or will develop torticollis related plagiocephaly (1 in 300 newborns). (http://www.ncbi.nlm.nih.gov/)
- Congenital muscular torticollis may be visible at birth or it may not become evident until several weeks later. The following are the most common symptoms of congenital muscular torticollis. However, each child may experience symptoms differently. Symptoms may include: tilting of the infant's head to one side, the infant's chin turns toward the opposite side, and firm, small, one to two centimeter mass in the middle of the sternocleidomastoid muscle. The symptoms of congenital muscular torticollis may resemble other neck masses or medical problems. Always consult your child's physician for a diagnosis. (http://www. healthsystem. virginia.edu/uvahealth/peds_ent/wryneck.cfm)
- Congenital muscular torticollis usually improves with range of motion and stretching exercises and massage, but it can lead to positional plagiocephaly and facial asymmetry if your child's head lies in the same position all of the time. In cases that aren't improving by 12–18 months, a surgical release/ lengthening of the sternoidcleidomastoid muscle may be required.

As I continued to research the issue on line, and as I continued to talk to these two moms and other who joined in on the topic, I became, in a very short period of time, rather smart on the topic—or at least smart enough to know what to ask and say to the pediatrician.

We returned to our pediatrician because we wanted more information. Frankie seemed to fit all of the classic symptoms. Our pediatrician listened to our story, examined Frankie again, and said that she would refer us to a pediatric neurologist. Two weeks later, and a visit to the pediatric neurologist confirmed this diagnosis for my daughter.

Fortunately for Frankie, her torticollis was diagnosed early, and we were able to do corrective physical therapy allowing the body to heal itself. While she may never have developed plagiocephaly or required muscle releasing surgery, many late diagnosed cases can develop these complications.

Another piece of information on torticollis I found rather interesting was that some information websites were claiming that torticollis was becoming more common because of the "back to sleep" movement, or getting mothers to put their babies to sleep on their backs to prevent Sudden Infant Death Syndrome.

> Why is torticollis becoming so common? Because babies are sleeping on their backs now and the muscles in the neck and upper back aren't being stretched out as they were with tummy sleeping. If you suspect your child may have torticollis or a tightening of the neck muscles, tummy time WHILE AWAKE, is vital to their recovery. Try to put your baby on their tummy whenever they are awake—at least an hour a day. (http://www.torticolliskids.org/)

Frankie actually preferred to sleep on her tummy, and I had been the dutiful mother waking up to put her back on her back every night, numerous times. In fact, in my sleep-deprived state of having two young children and continuing to try to be a full time academic, I could not seem to count as high as the number of times I had woken up only to switch my daughter from her stomach to her back—as she seemed to switch from her back to her stomach herself in her sleep. Sleeping on the back for babies is necessary to prevent SIDS, but according to torticolliskids.org, a parent information-sharing group, the lack of "tummy time" can contribute to torticollis in infants. I could not find these same statements on the health education websites that were sponsored by medical organizations, like the Mayo Clinic or the National Institutes of Health, and that served as a source of concern for me.

I brought this up with my daughter's pediatric neurologist and he insisted that back sleeping was crucial to my daughter's well-being but extensive daytime "tummy time" was essential. My own mother told me that I had been a stomach sleeper, too, and back in those days mothers were taught that stomach sleeping was a good thing. She also said, but that was before they knew what caused SIDS, so maybe you should listen to the doctor! What did the mothers think? Here their stories became interesting, and rather contested. A couple of moms thought that stomach sleeping was fine. They had a child that preferred that, and they allowed them to sleep that way, and their babies came to no harm. Other mothers insisted that daytime tummy time should be sufficient. The thread ultimately turned into a testimonial of what moms did and why they did it, including some detailed explanations of what some moms do to keep their baby happy during daytime tummy time.

I suppose that these points about whether babies should sleep on their backs or stomachs diverges a bit from my discussion of torticollis, however it was part of the same larger conversation I was part of as I tried to make my own best decisions about Frankie's health. It was again, however, experience bumping up against medical standards.

On Counterstories

It seems to me that my story about getting proper early intervention for Frankie's torticollis is a story about gaining access to the "right" kinds of information and the "right" kinds of doctors. As a mother I intuitively felt like Frankie's head tilting was abnormal—or at least it was rather different from how my older daughter's young neck and head tilted and bobbed. But how does a mother communicate such concerns to her daughter's doctor? Hello doctor, my daughter's head tilts too much to the right? Is that normal? How many times, as mothers, are we told that we worry too much about our children, and that these are normal experiences for the very young? How do you know when to persist?

What I appreciate about the mother's group I belong to is that many of the mothers responded to my initial query with their own stories. No one tried to give me the answer—nor was I seeking a right answer. I was seeking knowledge through experience. What had these other mothers seen, encountered, or heard stories of? What kind of knowledge and experiences could I arm myself with for a more thoughtful web search and conversation with my child's doctor? I wanted some evidence, grounded in a mother's experience, about whether, when, and how their children's heads tilted. Even though I ultimately gained most of my information about torticollis on the Internet, this conversation reminded me that claims related to children's physical development are always uncertain and child development varies greatly from child to child.

If we return to standpoint theory to make sense of my story about Frankie's torticollis, then I think it begins to help us to see how mothers' stories, which sit at the margin of the health community, serve as a foundational and situated web of ideas that give us key understandings and strategies for moving forward. I referred earlier to the importance of counterstories in allowing us to imagine a world beyond normative claims. In some respects, because mothers' stories sit at the margin of the health community and are not often part of the official body of science, their stories are counterstories. They expose, analyze, and challenge the stories of those in power, which are often the dominant discourse (Delgado, 1995). I do not deny I needed the knowledge and the power of doctors to help my daughter's medical situation. However, I learned from the moms in my on-line community that some children can express plagiocephaly sooner than normal development patterns predict. I did not want to wait until Frankie's six-month check-up to learn that her developmental patterns fell outside the normal bell curve.

Counterstorytelling can also build community among those at the margins and can show new and different possibilities by combining elements of the counterstory and the current reality. I believe that my queries about torticollis, and all of the other mothers' queries and stories about their experiences, have helped us to create a community where we can seek information when our own run-ins with the medical world leave us with unanswered questions. Standpoint theory—through counterstory—also helps us to see first hand the subjectivity of truth-making and to really re-position the meaning of epistemic privilege within and across communities. As a science educator I know that medical models are just that: models. Models do not perfectly describe each and every situation, and to know how or when to challenge those models to seek further medical advice is challenging, at best. Frankie's torticollis did not re-invent the world of torticollis research or treatment. She had a rather typical case with unproblematic treatment and recovery. But, being a mother who had to work hard to get serious consideration of her child does call epistemic authority and privilege into question.

Concluding Thoughts

Persisting in asking for a specialist analysis of Frankie's head tilting by a pediatric neurologist allowed her to get treatment early enough to prevent the possibility of plagiocephaly. But what if the odds were different or access to critical knowledge less readily available? What if the medical or insurance worlds persisted more strongly than me? Would have I had enough confidence and the right resources to pursue this in a way that was productive for my child?

Confronting and acting upon Frankie's torticollis raises questions about what it means to be scientifically literate I certainly did not learn in either my K-12 or university education about torticollis or even how to approach the medical community. Conceptions of science literacy must move beyond what one must "be able to know and do" to account for how such knowing and doing intertwine epistemological, methodological, and political activity. Scientific literacy should involve learning about the big ideas in science and about the nature of scientific knowledge, but this should take place in ways that also account for the personal process of coming to know in a highly contested world.

Note

1 Standpoint theory does not mention the use of counterstory, in so far as I have read. Counterstory is an idea that emerges from critical race theory, and has been defined by Delgado (1995) as the telling of stories of and by people whose experiences are not often told, such as low-income African American and Latino young people in urban schools. Counterstorytelling can serve as a tool for exposing, analyzing, and challenging the stories of those in power, which are often a part of dominant discourse. Counterstorytelling can build community among those at the margins, challenge the perceived wisdom of those at society's center by providing a context to transform established belief systems, and can show new and different possibilities by combining elements of the story and the current reality. I use the phrase intentionally here to suggest that the stories and experiences of those from marginal position are the most important stories in revealing the subjectivity of truth-making

and making problematic the kinds of epistemic privilege bestowed upon those with traditional access to scientific knowledge and production.

References

American Association for the Advancement of Science (AAAS). (1990). *Science for all Americans: Project 2061*. Washington, DC: Author.

American Association for the Advancement of Science (AAAS). (1993). *Benchmarks for science literacy*. New York: Oxford University Press.

Brown, B., Reveles, J., & Kelly, G. (2005). Science literacy and discursive identity: a theoretical framework for understanding science learning. *Science Education, 89*, 779–802.

Delgado, R. (1995). *Critical race theory: The cutting edge*. Philadelphia, PA: Temple University Press.

Harding, S. (2004a). A socially relevant philosophy of science? resources from standpoint theory's controversiality. *Hypatia, 19*, 25–47.

Harding, S. (2004b). *Standpoint reader*. New York: Routledge.

National Research Council (1996). *National science education standards*. Washington, DC: National Academy Press.

Roth, W.-M., & Lee, S. (2004). Science education as/for participation in the community. *Science Education, 88*, 263–291.

Wylie, A. (2003). Why standpoint matters. In R. Figueroa & S. Harding (Eds.), *Science and other cultures: Issues in philosophies of science and technology* (pp. 339–351). New York: Routledge.

9 Living with Chronic Illness

An Institutional Ethnography of (Medical) Science and Scientific Literacy in Everyday Life

Wolff-Michael Roth

It has been creeping up on me slowly. I feel increasingly tired but without really noticing that there is an increase. Walking up the stairs from my basement to the main floor fatigues my legs, and so does riding up the slightest incline on my bicycle. Then, in the summer of 2002, I repeatedly find myself confronted with an inability to make even minor decisions. In the first instance, I am standing in my kitchen about 10 feet from the couch in the family room. Feeling very tired, I thought I should lie down. But I cannot decide to walk the 10 feet to the couch. For a moment I am wrestling with the thought of what to do. And then, I simply let myself drop onto the slate floor to lie down where I instantly fall asleep.

A second, similar situation occurs prior to one of the lessons of the course I am teaching that summer. I am in my office preparing. And again, I simply lie down right where I am, in the middle of the office, and sleep. In both situations, I not only cannot make a decision, but also I realize that this is happening to me. I am unable to decide and realize that I am unable to decide, incapable of doing anything about it. It is a situation of an unavoidable radical passivity: I am struck with this severe fatigue that I have not wanted, with an incapacity that I can do little about. Through the experience of passivity with respect to illness, I come to understand something about all phenomena and learning more generally: "This powerlessness to stage the phenomenon, which compels us to await it and be vigilant, can be understood as our abandoning the decisive role in appearing to the phenomenon itself" (Marion, 2002, p. 132). That is, we are subject to phenomena as these are given (give themselves) to us, including those that, in illness, reflexively concern our experiencing experienced bodies themselves. The following copy of a note written during the month of July while teaching the course reflects the situation I have been finding myself in. At the time, I was also thinking a lot about experience and cognition from the perspective of the experiencing person, so, like many notes that summer, this one too, has the classifier "phenomenology," here with the sub-classifier "illness":

July 5, 2002 Phenomenology of Illness

Experience of illness completely orthogonal/detached from medical description

- iron deficiency
- symptoms

Even reading symptoms, their written form, [seems] completely detached from the summation of numbers I experienced at 8 a.m. sitting at my desk, trying to cope with email. "Lack of energy," "weakness" while trying to ride the bicycle so [as] "not to lose physical condition[,]" the pain/discomfort associated with it.

The description of "loss of focus" or problems of concentrating as the experience of struggling with reading, or even trying to compose an email.

The description [bore] no relation to the experience, any other could have done. For lack of better description, and [thinking] they would better understand medical language, this is what I gave to people without believing it myself.

This note shows how, long before I came to read critiques of (social) scientific concepts as ideologies that are introduced and become part of everyday life (Smith, 1990), I had become critical of the ways in which personal experience comes to be replaced by conceptual terms. These terms then take on their own lives, mediating the lives of people, such as when a student once labeled "learning disabled" comes to be jerked out of his regular classroom and social context "to be treated," though there is ample evidence that he is very literate both socially and scientifically (Roth & Barton, 2004). The ways in which my family physician and I can talk about what is going on with me become reified in notes that he enters in my file, and which may be transmitted to other medical offices in the case of transferals. Moreover, I have little influence on which terms are used to describe me and my condition.

The (what turns out to be mysterious) illness also mediates the ways in which I become attentive to the relation between my body, which is given to me in radical passivity, and my thinking about Self, which is no less given in radical passivity. I realize how particulars of the language I speak and write separate who I am from my (material, emotional, experiencing, suffering) body. Nearly three weeks later, I make the following entry into my research notebook:

July 25, 2002 Phenomenology of Illness

"I" am running into a brick wall
- presence and absence of "I"
- happening psychically and psycho-emotionally
- tendency to say "my" body, but what is it apart from "I"
- "I" is also my body, refers to me as "person," . . .

The note points to the seriousness of the situation and to the fact that in all its mysteriousness, the illness is touching me to the core. The descriptions I use include "running into a brick wall" and, not noted here, "running on empty." Over time, I begin to experience a variety of other conditions. My knee, elbow, hip, and upper-arm joints begin to hurt as if nerves were pinched in these places. I do know about the abrasions of cartilage at least in one of my hips, as I had been diagnosed with pre-arthrosis (an articular illness that produces fibrosis or degeneration of the cartilage) as a teenager. I experience sudden loss of blood and numbness in my fingers, which turn completely white, a condition that I later learn to be "Raynaud's syndrome."

Although I have never been eager to spend (waste) time seeing doctors—spending time to do the work of waiting (Smith, 2005)—and have never exhibited hypochondriac tendencies, I have begun going to see my family physician. I decide to see him in part because of the strong encouragement my wife Sylvie provides in doing something about my situation: "You can't go on like this. You have to do something." And when I do see the physician and specialists, I find that the medical practitioners specifically, and the medical sciences more generally, are as much bricoleurs tinkering toward success as are the scientists and other professionals I have been both studying myself and reading about in social studies of science literature.

At the time, such reflections encourage me to return to a project that I have let slip from view, the experience of learning from the position of the learner, so that, in due course, I write *Learning science: A singular plural perspective* (Roth, 2006). But many of the notes concerning the phenomenology of knowing, learning, and experience that ultimately led to writing that book have not entered it, despite their apparent mediational nature of bringing it about. This auto/biographical connection is important to my work because it brought about much of what I have done since and what I have published in a variety of forms; and this auto/biographical experience led me to a substantial rethinking of agency in terms of radical passivity and givenness, the fact that there are many aspects in life generally and in learning (science) specifically that are not made thematic in current theories of learning—including the fact that we do not intend our intentions to learn and the fact that what we learn has been beyond all reach just prior to the moment that it has become salient to us. Thus, both the intention to learn and the object of learning are given to us, or rather, give themselves to us. And we, the learners, are but willing hosts—and even hostages—accepting in radical passivity what has come to us.

Despite many science educators' claims to the contrary, I see little if any evidence that we need to know any scientific knowledge and theories to do (financially) well in and lead a happy life, or to contribute in an ethico-moral way to society by "aiming at the true [good] life with and for others in just institutions" (Ricœur, 1990, p. 211, my translation). In fact, the progressive cultural-historical division of labor of society and the increasingly expanding nature of specialized knowledges makes it nearly impossible to learn all that we need in everyday life. And under what criteria should we select particular elements for the curriculum? Why, for example, should it be more important to learn Newton's third law and not how to repair small engines such as those found in lawnmowers, leaf blowers, and weed eaters? Conversely, there is a need for competencies in critical literacy, as exemplified in my experiences with the medical sciences and medical practice during a mysterious and chronic illness. In this chapter, I use these experiences, and medical sciences and medical practice as the place from which to conduct an institutional ethnography for the purpose of developing a critical stance towards science and scientific literacy where it is needed most: right in my body and in my life. I describe my coming to know the medical sciences generally and the medical system in particular, my eventual search for solutions in those moments when I felt I was not getting any help to analyze and reflect on the form(s) of (scientific) literacy that I might want for people like myself, who find themselves in situations that they want to get themselves out of. I conclude with reflections on the kind of (scientific) literacy I would want for patients, doctors, and others spared (chronic) illness.

Learning Medical Facts and Learning about the Medical System

Once I place myself in the hands of doctors specifically and in the medical system more generally, something like a hidden machine begins to operate and I no longer have options to intervene: I become a willing subject/object. The physician sends me to this or that specialist, who returns results that I never see to the physician who reads and interprets them for me—much like the Catholic Church used to keep the Bible in Latin so that ordinary people could not read it. Although I may be considered to be scientifically literate far beyond the norm in society, I am kept from accessing the documents produced and from participating in interpreting them and relating them to my experiences.

There also is a temporality in the process that is difficult to bear. Because "causes have to be eliminated one at a time" (physician), there is no way in which a more comprehensive approach to my utterly real problems becomes possible. Each specialist attends to me in the way that he or she knows best without consideration of the larger context in which I am embedded and without consideration of what all the other specialists have done and found out. I am experiencing in and with my own body how the "one-variable approach" typical of the "scientific method" is failing me.[1] And frequently I learn that the physician, who is in the position of coordinating the investigation into my chronic illness as a whole, does not have the willingness or the background to pull the divergent analyses into a more coherent picture. As a real-life person, I experience everyday medical practice, which grounds itself in medical sciences, and learn the facts it attends to both in my interactions with the system and from what I learn while perusing the Internet and reading studies published in the medical journals that I access through my university library. Furthermore, much like the scientists and engineers in my village, who do not pay attention to and disregard the extensive 30-year historical knowledge that local residents had built up while living in their homes in decision-making about the ecology of their water supply—instead trusting the one-shot measurements a single engineer had conducted—the doctors examining me show disattention to, and disregard for, my extended knowledge about my everyday life and context.

Family Physician

In Canada, where we have a universal, accessible-to-all healthcare system, the family physician is the first point of call for any person needing medical care. If needed, the physician then refers the patient to a specialist, depending on the assessment of the symptoms. In my experience with the medical system, there appears to be a tendency to minimize a patient's account, perhaps especially in cases such as mine when the patient appears to be fine. Thus, my physician orders a "complete" blood test, which turns out normal results for every test requested, among others for hemoglobin—low levels of which would indicate anemia, a condition infrequent in developed countries but rampant in tropical countries (Mol & Law, 1994). But my fatigue persists. A first indication of problems emerges when I happen to find out that there are forms of anemia that are unrelated to the amount of iron in the blood. Before he dies that year, my father-in-law, who had practiced family medicine in northern France for more than three decades, suggests I should get my iron tests. I tell him that this has been done, but he insists, no, it is not blood iron that needs to be tested but the iron storage, the ferritin levels. Serum ferritin is a globular protein complex that consists of 24 protein subunits. It is the main intracellular iron storage protein in both prokaryotes and eukaryotes. My Internet search reveals that a ferritin test is the "best indicator" of iron deficiency. I return to my family physician and ask him for a test of ferritin: it turns out that with a (for this lab) normal, reference range from 40 to 300, my level is 8. When the physician tells me the results of the test, it becomes evident to me that the situation is serious; and the low levels are a highly likely cause of my fatigue. But in all the seriousness the numbers invoke, they stand in no relation to my experience of fatigue. As my note dated July 5, 2002 shows, the medical discourse—technical words, the numbers the medical system produced, the physician's explanations—seemed to have no bearing on what is happening to me. The clinical gaze produces numbers that—like artifacts more generally (Winner, 1980)—produce and reproduce a particular politics associated with them. In another note, I describe the relation between the two: the discourse describing my situation and the medical explications appear to be "orthogonal to," on a different plane than, my daily experience of fatigue.

The physician recommends a daily supplement of a particular form of ferritin and asks me to go for another blood test two months down the road and to return for another consultation. During the period, I take my daily dose of ferrous gluconate. When I see my physician the next time, however, the serum ferritin level has increased by just one point, to 9, and remains far from the lower limit of 40. The physician recommends doubling the dose, to get another test two months down the road and to return. This time, my score is still below

20 so that the physician recommends tripling the dose and to go through the same process of testing and returning for a consultation.

My extended fatigue and the problems with concentrating and writing begin to raise my anxiety levels. I think about future evaluations, those for salary purposes in my department and faculty and those for grant purposes in my national funding council. Although I describe the situation to the physician, I have to make a special effort to convince him to write a note as to the effect of the continued fatigue. He makes me pay $10 for a simple note (Figure 9.1). The note itself exhibits little of what is happening to me. The fact that the physician does not consider it important that my workplace be informed about my continued medical problems shows the low level of attention that this physician particularly and the medical system generally pays to me and other patients. This, unfortunately, is not a singular event, as I find out while I am writing this text in the summer of 2007. After dinner, I first throw up and then begin to have cramps of sufficient magnitude that I seek medical attention in a walk-in clinic. Again, the treating physician says, "It is nothing," though I insist that I am in terrible pain. He then continues, "If the problems persist come back tomorrow." Although my description includes the possibility of having eaten a potato with a green splotch—which I know leads to solanine poisoning that in serious cases lead to death—he disattends to what I am saying. When it is evident that he does not even know what I am talking about ("What is this?"), I explain it and the symptoms of mild poisoning, which precisely are those I exhibit. Here, my knowledge, augmented with medical facts and conditions that I have accessed through the Internet, is disregarded rather than taken into account. Again, I learn about (medical) science in everyday life and how it is treated as an aspect of but not as a resource for life.

SEPTEMBER 16, 2002

TO WHOM IT MAY CONCERN:

RE: W MICHAEL ROTH
DOB: JUNE 28, 1953

This man was disabled June to October 2001 because of a cycling accident.

He has also been partly disabled by illness from May 2002 to the present. His current illness is ongoing.

YOURS SINCERELY

[[signature]]

Figure 9.1 Fearing lower scholarly productivity and the repercussions this may have on my salary increments and funding requests, I ask the physician for a written statement that I could use for future reference.

In the course of the investigation about possible sources/causes of my ever-extending illness, I see numerous specialists, including several neurologists, a rheumatologist, and an allergist, each time preceded and followed by one or more visits to the family physician. The topology of my illness is extended, including a variety of similar and dissimilar regions, connected by a flow consisting of documents and my body. I remember asking myself whether we, the documents and I, are the immutable mobiles of social topologies and network theory? Or is my body a boundary object that leads to very different practices in the different regions of the network it brings about—duly represented in my medical record?

Neurology

Some of the symptoms I experience include tingling in the legs, cramps, muscle weakness, and continuous muscular activity (twitching). The physician sends me to the one of the neurologists I had seen after my bicycle accident in 2001, with the subsequent identification of damage to the nerve paths in the brain or spinal cord as evidenced in the Babinski reflex. (The Babinski reflex occurs when the great toe flexes upward while the other toes fan out after the sole of the foot has been firmly stroked. This is normal in young children until the age of 2 but a sign of nerve path damage in individuals older than that.) As during the months following my bicycle accident, the neurologist detects problems with the transmission of electric signals in my right leg and shows concern at the constant flickering (twitching) of muscles in both legs.

The neurologist conducts an intake interview without, however, asking more about my life generally, my eating habits, or the intensity and frequency with which I exercise. I am very interested in some of the tests, one of which measures the rate at which electrical pulses travel a measured distance from the thigh to the ankle. This test reminds me a lot of those that I had conducted pre- and post-graduation with my MSc degree in physics, when I had worked for an advanced research lab building probes for measuring the rate at which heat pulses travel in healthy and diseased gum tissue. As an anthropologist of science, I am also interested in her activity watching the use and transformation of my body into a variety of inscriptions, both on computer monitor and in print-outs. I feel that I understand, at least in principle, what she was doing and why she was doing it.

She refers me to another neurologist in the same office but with different specialization, who, following additional tests and consultations with his colleague, diagnoses a high probability of neuromuscular disease and a 5% probability of amyotrophic lateral sclerosis (ALS), which in North America also is called Lou Gehrig's disease. Being only vaguely familiar with the term, I ask him what the consequences of ALS are and request him to be open with me. He says that it is associated with an average lifespan of three years and a progressive weakening of muscles in the body, including those in the chest that support breathing, and eventual death. I remember being afraid, not for myself, as I had decided long ago to lead my life so that I could depart any day, but for my wife, who would have to deal with a loss.

Both neurologists agree, however, that I should see the foremost specialist in our province and they promise to line up a consultation, which, given the seriousness of the diagnosis, takes place only 10 weeks later. In the meantime, I become a specialist concerning different forms of muscular dystrophy. Most importantly, all our planning for the future comes to a halt because we do not know the final diagnosis. We think about wheelchair access to the home and how we might deal with the rapidly progressing degeneration. I find out that the husband of a university colleague has just been diagnosed with ALS, and she shares with me her experiences of the rapid incapacitation that has been occurring. Here, the topology of my illness begins to include, for me at least, others affected by the illness that looms above my head, though this part of my network is of no relevance to the medical system. She tells me that she does not think I have the disease, despite the assertions of the neurologists; but I take her comments with a grain of salt, as there is always the possibility that she wants to comfort me. In the meantime, nothing else happens. The physician explains again that "we have to exclude one possible cause at a time"—especially after I suggest that a considerable number of my symptoms map onto those of chronic fatigue syndrome.

Eventually I see the top neurologist, who repeats the same tests I have undergone before—though this is not a mathematical relation of identity A = A. He eventually says, "I do see why other neurologists have made the diagnosis, but based on my extensive investigation, it is not likely that you have ALS or another neuromuscular disease. There is considerable muscular noise in your leg, but it is probably due to other causes than those underlying dystrophy." My wife, who accompanies me on the trip, and I are so relieved that we buy a few very good bottles of wine on our three-and-a-half-hour journey home. The announcement has come as a big relief, though the end of coming to understand what is happening to me is not in sight.

Rheumatology

My symptoms persist, especially the pain in several joints, including the hip and knees. When I see the rheumatologist, he seems very skeptical of my account, which includes a description of all the various symptoms I have experienced—including the one about the curious ammonia smell after exercise. I have the firm belief that the issue is one of identifying possible causes for my (mysterious) illness and that a complete description of symptoms and context is required—there are conjectures that conditions such as anemia are not to be found solely in the body but within sociomaterial systems that have a fluid topology (Mol & Law, 1994). I strongly remember the sense of being looked upon as a hypochondriac, especially when the rheumatologist asks me to come to the point and comments on some descriptions as not pertaining to the issue—for example, the ammonia odor during and after exercise.

Despite his apparent skepticism, the rheumatologist conducts a series of tests, including reflexes, and investigates the joints. He orders a series of x-rays to be done and sends me home. I find out about the test results from my physician, who tells me—without allowing me to see and make sense of them—that the x-ray images show that there is absolutely nothing wrong with me. All my joints are well and healthy with no degenerative effects to be seen. But I am also thinking that, with all of the tests, nobody has yet looked at me as a whole person-in-context rather than at some symptoms that fall into the experiential domain of a particular doctor. "I" am reduced to a set of x-ray images showing my hip and knee joints from the front and side. Although I was taken care of by a medical system, it behaved like an assemblage of independent elements rather than as a system where the moments are mutually constitutive. At that time, in my work, I was looking at activity theory as a way of understanding, among others, medical systems. I see that I have become something like a boundary object that is moved between the different nodes of a system, understood in terms of the particular node irrespective of the findings of other nodes. And there is no instance where the different forms of knowledge associated with my body and file would be collated, explicated, and understood in a holistic way. More specifically, there is no mechanism that would allow my extended historical (autobiographical) knowledge to enter and mediate interpretations; and I am excluded, too, from the interpretive process.

Allergology and Immunology

In May 2004, I have an appointment with the allergist. Like the other specialists before, he asks me about my problems but shows little interest in any of the context. He does not ask questions about what I eat or how much I eat; he does not ask about the amount and intensity of exercise I do. All he is interested in is the state of fatigue and he tells me that his assistant will conduct a test for a variety of common allergens. I observe the events with great interest, and use my presence here as an occasion to (informally) study medical practices, including production and transformation phenomena and inscriptions (Figures 9.2, 9.3). I decide to follow inscriptions and their translations, as suggested in a well-known primer on science in action and its advice on how to follow scientists and engineers through society (Latour, 1987).

The assistant uses a pen to mark an array of 32 points on my left forearm (Figure 9.2), arranged in 4 rows of 8 points. She marks two additional points outside the array, which she explains, upon my query, saying that they are reference points where no allergen will be placed. I think, "Ah, these are the controls." She then brings a tray with different vials (Figure 9.3, left), places a drop of liquid from each on one of the pen marks (Figure 9.3, right), checking on a sheet of paper those positions on my arm that already have received an allergen (Figure 9.3, center). She asks me to wait for 10–15 minutes until the doctor comes back to take a look. While I wait, I can see that there are areas around two of the points where my skin begins to redden (the third and fourth

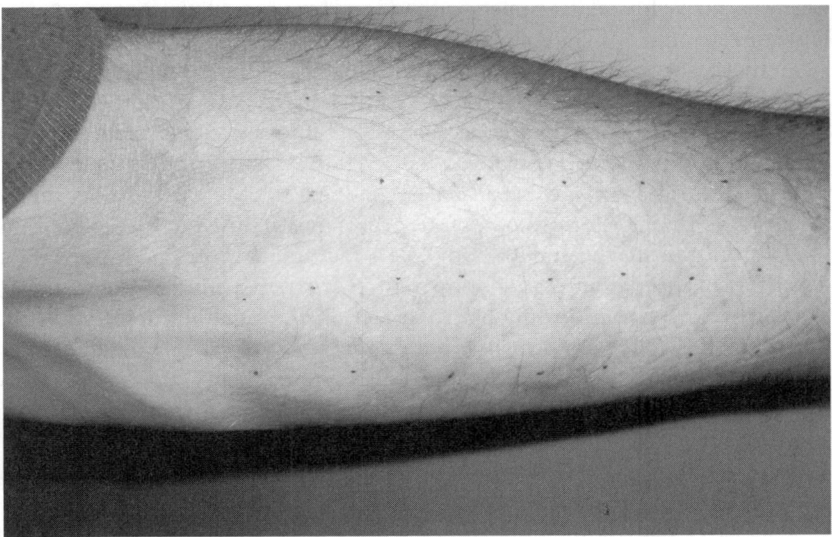

Figure 9.2 The allergist conducts a test for 32 common allergens on a grid marked on my left forearm, and includes two reference points ("controls") that do not receive allergens (far left).

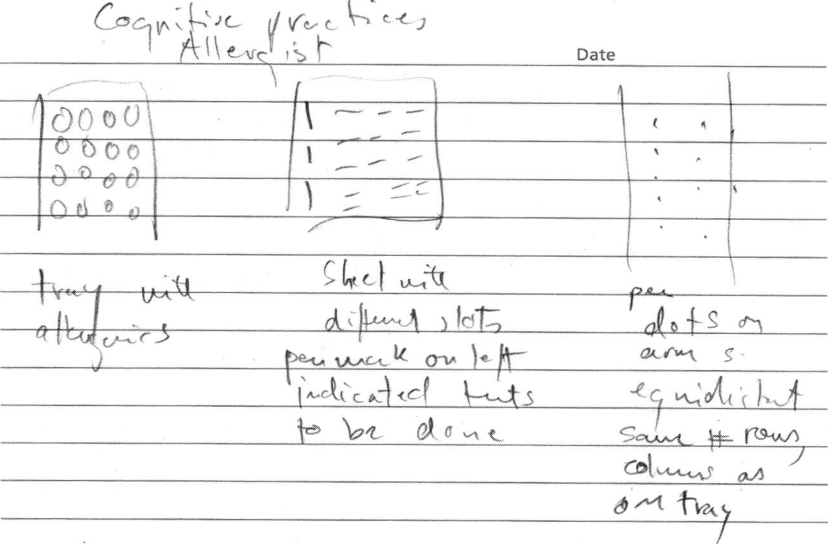

Figure 9.3 Some of my ethnographic notes taken right after the visit to the allergologist concerning the cognitive practices encountered there.

from the left in the bottom row). As an ethnographer, here of the medical sciences, I have my digital camera with me and take several photographs before anyone returns.

In this particular situation, as the anthropologist of the medical sciences that I am in this instance, I understand myself to be translated again, this time not into the x-ray images of the rheumatologist but into a set of representations on paper and on my arm (Figures 9.2, 9.3). While I wait for the allergologist to return, I think of one study in particular, conducted by Marc Berg (1997) on the use of inscriptions—forms, medical records—that are used to model both the patient (or rather, the patient's body), here me, and the work performed, here clearly visible in the translations apparent in Figure 9.3 and the (momentary) traces the work leaves on my body. In fact, in the situation depicted in the photograph (Figure 9.2), my whole body comes to be represented in the 34 pen dots and the lack or absence of reactions surrounding them—a part of my body comes to stand for its entirety and its problem(s) in a metonymic form. The two reactions on my skin allow my classification into the allergy slots of "dust" and "mite." But these slots retain none of my joint pain, none of my weight loss, none of my problems with concentrating (reading) and remembering.

This (my) body is the empirical, the informal to be disciplined through formalization. The forms and inscriptions are ways of formalizing my body and thereby producing formal medical knowledge that can be compared against the standard lexicon of medical conditions. Here again, formal knowledge does not appear to map onto what I know about my condition, its history and historical development, and what is happening to me at the moment. In particular, the forms do not have any slots for recording the history of my illness, which is only available in and through my narrative accounts. Formalization abstracts (detaches) from specific localities without, so the underlying ideology goes, losing its essence. Yet if the essence of an illness includes the setting and context besides the human body, formalization as it occurs in the offices and laboratories of the neurologists, rheumatologist, and allergologist is unable to provide the correct explanation for the symptoms and their causes. It appears to me that the methods of (medical) science are precisely failing where they ought to help; in formalization they become blind to and exclude the very knowledge that might mediate a viable interpretation leading to an understanding of what is happening to me. The specialists rely on the forms and formalized knowledge to render their judgment without querying whether their maps correspond to what is happening in and to my body, the territory described by their maps—despite the well-known assertion and caution, ascribed to Alfred Korzybski, that the map is not the territory. Between the two, the map and the territory, translations are performed, and it is beginning to occur to me that "I," the territory, get lost in the translation.

Upon returning to my examination cubicle and upon inspecting the grid, the allergist explains that I am allergic to dust and mites. Both are commonly found in (wall-to-wall) carpets—which I tell him to have eliminated entirely from my home associating this with a decrease in my chronic sinusitis, with which I have been afflicted for more than 40 years—and in beds. He asks me whether I have experienced sleep apnea and whether I snore, both of which I answer negatively. Yet he continues to explain that the inevitable mites might cause a swelling of the sinuses and therefore problems in breathing, which would lead to poor sleep and fatigue. He recommends special sheets and covers for pillows and comforter, liquids to clean my nasal passages with, and some other implements to be used, some of which are available in his office at the front desk.

In this medical office, my body is translated onto a grid, which makes visible a certain aspect of it. But the process is not unlike the one that Procrustes enacted, when he stretched or amputated his guests so that they would fit perfectly his (secretly adjustable) bed. Where the gap between the informal, my body, and the formal, the inscriptions that the allergologist works from, has to be crossed, the one person who could be called as "expert witness" is left out. What I say—and this despite the possible credibility my science degrees could confer to my status as witness—bears little on the decision-making and practical recommendations produced here or elsewhere in the medical practice charged with finding out about my illness.

Conclusions of the Standard Medical System

In the end, my family physician tells me that standard medical system has no answers for my problems; or rather, it has eliminated some of the possible causes of my illness. He summarizes the findings of the different specialists: there is no neurological problem to explain muscle fatigue, nothing wrong with my joints that would explain my joint pain, and my allergies are too common to do anything other than what the allergist recommended. He says that his best guess is this diagnosis: "chronic fatigue syndrome slash fibromyalgia." I am not surprised, because my own research on the Internet has revealed that many of the symptoms I exhibit are associated with the two illnesses. Disturbingly to me, the physician says that there is nothing he can do, that there is no form of medical help—though I later find out that there exist clinics, even quite near where I live, that specialize in assisting people dealing with chronic fatigue syndrome and fibromyalgia.

Throughout my chronic illness, which is abating but has lasted to this day, it is rather astonishing that chronic fatigue syndrome (CFS) has not been the (definitive) diagnosis. I do remember the physician stating—when I ask him to check out CFS—that "chronic fatigue syndrome is the last thing we check out when nothing else shows up positive." Yet the definition of the illness provided on the Internet by the U.S. Center for Disease Control squarely fits the symptoms I have been exhibiting for a long while (CDC, 2006):

- have severe chronic fatigue of six months or longer duration with other known medical conditions excluded by clinical diagnosis; and
- concurrently have four or more of the following symptoms: substantial impairment in short-term memory or concentration; sore throat; tender

> lymph nodes; muscle pain; multi-joint pain without swelling or redness; headaches of a new type, pattern or severity; unrefreshing sleep; and post-exertional malaise lasting more than 24 hours.

But the same website also states that "health care professionals may hesitate to give patients a diagnosis of CFS for various reasons." At the same time it recognizes that it is "important to receive an appropriate and accurate diagnosis to guide treatment and further evaluation."

In all of this, I never see a single result or report. Perhaps I have the right to see them, but nobody offers me to let me read them personally. The doctors are like medieval priests with their Bible: Of what help is my scientific literacy? It is always the physician who reads from a sheet, which he holds like a card player tight to his chest preventing others from seeing his hand. All the specialists and my physicians know that I am a professor at the university studying the knowing and learning of science, and that I am a trained scientist. Yet nobody ever seems to care to explain a process or product of the ongoing investigation. At the very moment that I am in the medical system, experiencing its fallibility, the various staff act in ways that reproduce the very myths of medical science that make them something special in society. I experience and come to understand medical practitioners, specialists as well as generalists, as bricoleurs who often tinker toward success but not this time. This is not problematic for me, as I have come to study different scientific and technical professions, where much of the knowledge is tacit, unarticulated, and radically pragmatic (rather than conceptual, theoretical). What is beginning to bother me is the apparent pretense of medicine to be a more or less exact science when all the while it is apparent that it is failing to diagnose my problems.

In the end, my physician acknowledges the failure of standard medical knowledge and the medical system to arrive at a diagnosis. He suggests that I try alternative medicines. I do not know what he thinks himself—whether he believes that I am merely pretending, despite my repeated wish that the process could be quicker and a solution reached, and my being tired of sitting around in doctors' offices, waiting for appointments.

Further (Parallel) Investigations

As a result of my continued illness, I begin my parallel investigations by observing what is happening to me more closely. One of the things I remark is a curious smell of ammonia after exercise, which I noted after taking off my all-weather jacket. But neither the physician nor the specialists I see show any interest in this observation, suggesting either that it is something completely unrelated or simply not reacting at all to this part of my account. One day, however, using "ammonia" and "exercise" as search terms, I come across an Internet site where this phenomenon is being discussed—in the absence of sufficient protein the body will begin using its own muscle protein and, in the process, produce ammonia. After talking to the owners and employees of my bicycle shop, who appear to be familiar with this phenomenon, I investigate creatine—2-(carbamimidoyl-methyl-amino)acetic acid, normally found in red meat—to use it as a dietary supplement. I am especially interested in creatine after reading in the medical literature that clinical trials have shown benefits in individuals with ALS, which is the hypothesis the medical system is pursuing at the time.

As I have done throughout the illness, I carry out extensive searches on the Internet concerning this supplement, which appears to be favored within the young sport community for increasing energy and lean muscle mass. However, despite the recommendations of the people in the bicycle store, and despite the apparently positive effects of creatine reported on many websites, I am more interested in the medical research. Although I am taking a critical and reflexive stance toward the field, I feel that looking up original studies on the use of creatine and its (side) effects would help me make a better decision. The recurrent problem with perusing the medical literature—which I easily access online through my university library—is my unfamiliarity with the discourses of the fields, which turn out to cover the sciences (e.g., cardiovascular medicine, chemistry, neurology, molecular biology, medical physics). I feel confronted with a continuously changing "gobbledygook" as I move from one article to the next.

I begin using the product as a dietary supplement and experience an almost instant decrease and elimination of the ammonia smell and a slight improvement of my muscular condition. But I eventually abandon taking creatine supplements. After the ALS scare is over, and having found that I am beginning to suffer from some of the side effects—e.g., dehydration and cramps—and fearing other listed side effects like renal stress and failure, I end this venture into creatine supplementation. I also learn to look out for mediating circumstances that change the effects of a particular supplement or medication. For example, I find out how different foods mediate the absorption of iron supplements. Coffee, tea, eggs, milk, and a high-fiber diet inhibit the absorption of the ferrous gluconate that I take in the early stages of my illness, but vitamin C enhances absorption. Although grapefruit has a lot of vitamin C, it also appears to attach to many of the binding sites medication and supplements normally use so that there are recommendations not to use this fruit as a source of vitamin C.

Alternative Medicine

I begin to explore alternative medicine more intensely after the regular medical system fails to identify what is happening to me and after my physician recommends it as a possible extension of present forms of diagnosis. Although I have been thinking about alternative forms of medicine, and although I have personal knowledge of situations where alternative forms of treatment helped not only to alleviate but also eliminate conditions that the medical system had failed to treat, I am not eager to seek help from practitioners whose disciplines, in many cases, are not based on formal studies of their methods and results. That is, although I have been critical of medical research, because it often confounds correlation with causation and because it misleads by articulating causal relations on statistical grounds that are far from pertaining to every case (unlike Newton's laws pertaining to objects in gravitational fields, which do apply to every case), I still trust standard medicine more than its alternative cousins. I am also reluctant to submit myself to practitioners whose work is not recognized by society in the sense that the costs involved have to be borne by the user rather than by the system and private medical insurance.

My hesitation over going to naturopathic doctors is linked to the classification of their practices as "alternative medicine." Like the conceptual change researcher's "alternative conceptions," the label "alternative medicine" has a ring of being the other (usually lesser) form of something. Does the term denote an alternative medicine or is it medicine practiced alternatively? The adjective "alternative" marks out territory, is an artifact in and part of a particular politics (Smith, 1999). This hesitation appears to be justified by the many less than genuine advertisements in the journals of the alternative medical field that sound more like advertisements by medieval quacks than something I want to rely on. There are, for example, "Ion cleansing detoxification footbaths" available for sale, devices that "restore the physical properties of spring water," "magnetic washballs" that save on detergent and help the environment, or a "negative ion air purification system with ultraviolet light" that "effectively removes pollutants down to a scant 0.01 microns in size." Moreover, next to the names of the authors in magazines devoted to alternative medicine and alternative approaches to health there are single acronyms, or lists of them, used as resources for constructing legitimacy: ND, DrTCM, RAc, CCH, RSHom (NA), RD, RNCP, CIH, PhD, BSc, Pharm.[2] As soon as readers lend more credence to an article because of such a label, they have accepted the power of the concept of a degree and have become part of the conceptual power of the field. Despite my hesitancy, mediated by the persistence and increasingly debilitating effects of the illness, I eventually turn to alternative medicine, without dropping my critical approach, however.

On what basis could I select a medical practitioner? If we consider regular and alternative doctors as practitioners (rather than researchers) of medical science, on what knowledge basis should I make an informed decision? How can I make an informed decision if I have no knowledge of the field? Several people at my wife's workplace recommend the first naturopathic doctor I eventually come to visit. His curriculum vitae is posted on the Internet and reveals work experience and education that, in part, fall within the "credible" domain. He has a Bachelor of Science degree from a local university, where he also completed pre-med studies and he has obtained a naturopathic doctorate (ND) from a college in the US. He has participated in biophysical and cancer research and taught microbiology at my university and a variety of courses in colleges of alternative medicine, including traditional Chinese medicine. He is registered with a Canadian organization accrediting naturopathic doctors.

My visit turns out to be very disappointing, especially given its cost of $100. He listens to my description of symptoms and then investigates my tongue and fingernails. After I tell him that I hold an MSc degree in physics and that I have done doctoral work in physical chemistry, he appears to be more cautious with his assertions about physical and chemical processes.

He suggests that there is likely something going on with the kidneys and liver, as I seem to have congested lines of Qi, and that a dose of vitamin B12—a "special" preparation of which I can buy at his front desk, where "special" denotes the fact that his formula is better absorbed than others on the market—will assist. If after taking it for two weeks I do not experience an improvement, then I can come back for another half-hour session (at $50 cost) for further consultation. He also offers me intramuscular shots of vitamin B12, which, in his words, would be better absorbed by the body.

After this experience, I feel like I have gone to two doctors, both my first and last naturopathic doctor. Other than that he has spent an entire hour with me, there was little to distinguish his investigation and diagnosis from conventional medicine. He rapidly comes to the conclusion that it is my liver, on the basis of traditional Chinese medical knowledge as indicated by my fingernails. He has not asked me about my way of life, the quantities and types of food I eat, the amount of exercise I get, or any other contextual information that would place me in the center of an open system that one needs to understand to know what the causes of my problems are. He does not ask me to keep a precise log of the foods I eat, their quantities, and or the events in my life. I am especially dissatisfied because the costs of alternative medicine are not covered by the provincial (public) and private medical insurances; the lack of knowledge in the system and on the part of the doctors is thereby unloaded onto the clients ("consumers"). Because he does not know—a fact he does not openly admit—I am asked to return for further investigation. I feel like I am paying so that he can learn and gather experience.

Although I know upon leaving that I will not come back, I am intrigued by his indication that I might suffer from a lack of vitamin B12. When I look it up on the Internet, I find that there is a form of anemia induced by the lack of this vitamin, which humans essentially get from the consumption of meat. Vegetarians often are deficient and need dietary supplements. Some of the symptoms of B12 deficiency map onto what I experience, such as weakness and fatigue, dizziness, poor appetite, and weight loss. In patients with prolonged B12 deficiency, the symptoms include muscle weakness and tingling, numbness, memory problems, loss of balance in walking, and irritability. Because I know about confirmatory bias, however, I especially look for symptoms that I do not have as possible counter-evidence. Thus, I note that I do not experience dementia, diarrhoea, depression, or psychosis. Interestingly, however, in the course of writing this chapter, I find out that a cause of vitamin B12 deficiency may be fish tapeworm, an infection of which can come from eating raw fish, especially salmon—sushi, sashimi, surface seared—which I regularly eat as it is a delicacy. This tapeworm competes for vitamin B12 and interferes with its absorption, leaving the host subject to pernicious and megaplastic anemia. That is, there is a form of anemia that has never been taken into consideration while interpreting my symptoms.

The experience with the second practitioner of naturopathic medicine turns out the same. I make a second attempt because a colleague strongly recommends the doctor. The colleague is a world-renowned scientist who regularly publishes articles in *Nature* and *Science*, and is member of the Order of Canada and other honorable societies. He shows up regularly on television providing a scientific perspective on certain global environmental problems. At the time he indicates to me that his high degree of skepticism with respect to alternative medicine was abated when one of his children received help from a naturopathic doctor after the regular medical system had failed them. Following that experience, his entire family found relief for a variety of medical problems they have had. As a result, he had come to trust naturopathic medicine, at least as this one doctor practiced it.

Deciding to give alternative medicine another try, I make an appointment with my colleague's naturopathic doctor. However, I become suspicious while I am going through the intake interview. Like other regular and naturopathic doctors, this one focuses on a small number of things and does not check what seems to me the most obvious set of questions relating to the type of life the patient is leading. My suspicion is increased when the doctor, among other things, is astonished by my gold crowns. He suggests that the metal causes an electrical potential with respect to the inside of the cheeks, which is a potential cause of my problems. I remember to this day my private thoughts about the low reactivity and solubility of gold, and my attempts to recall electrochemistry. As a physical chemist I know that there are electrical effects in the presence of another metal, which causes potential differences to occur when two different metals touch, and that I experience it each time I touch a tooth with a metal spoon or other metal object. Standard electrical potentials are calculated with respect to the recombination of free hydrogen ions in aqueous solution and electrons to form gaseous hydrogen according to the chemical equation

$$2H^{+}(aq) + 2e^{-} \rightarrow H_2(g). \qquad (9.1)$$

I think that unless there is another metal present or something in the skin that sets up an electrical potential, nothing should be happening. (I find out only later that there are reported cases of gold contact allergies, a T-cell mediated immunological state. But studies mainly pertain to contact allergies, for which a meta-analysis shows that the reactions turn up as stomatitis, that is, inflammations of the mucous lining of the mouth.)

In the office, while the secretary calculates the costs of the examinations that the doctor has recommended, I peruse the information available about the tests in a ring binder I am given while waiting. I take a few notes, including the names of conditions that the doctor wants to test me for. One of these is Candida albicans. In the meantime, the secretary has totaled the cost of the examinations: over $700, including about $150 for the Candida test and $60 for measuring the electrical potential between my gold crowns and cheeks.

Immediately after returning home, I check out Candida albicans by seeking information on the Internet. I take several tests where I answer questions such as: (a) Have you taken any antibiotics within the past year? (b) Do you presently have any of the following symptoms: athlete's foot, jock itch, or vaginitis? (c) Do you have a sore or burning tongue? and (d) Do you have bloating and/or intestinal gas? All of the tests indicate that there is little likelihood that Candida albicans causes my problems. This doctor wants something like $150 to do a test for the yeast. I later check with the dentist, who begins to chuckle when I tell her what the naturopathic doctor had said about the electrical potential and its eroding effect on my health.

In the end, I walk away from "alternative medicine" with suspicions similar to those I have toward the dominant medical system—there is little concern for the patient-in-context and when diagnosis is difficult, the lack of knowledge remains unacknowledged. It is true, I have contacted only two naturopathic doctors, but, because both were highly recommended, I tend to apply Bayesian inference, whereby the degree of belief in a hypothesis changes as evidence accumulates. In the field of alternative medicine, the lack of knowledge is unloaded onto the patient in the form of payments, which in fact constitute fees that allow doctors to gain experience and experiential knowledge. One naturopathic doctor has reduced the complexity of illness-related issues to what he can see on my fingernails—without telling me what it is he sees—thereby engaging in a metonymy-producing process similar to the allergologist; the other announces that identification requires multiple translations of my body into inscriptions of various forms, each translation also mapped onto a grid of fees (like a catalogue of goods where you check off the items you want to order).

Personal Research and the Praxis of Really Alternative Medicine

In the course of seeking information on the Internet to parallel my consultations with doctors in regular and alternative medicine, I find out a lot about chronic fatigue syndrome and fibromyalgia. I also come to know about cases at my wife's (Sylvie) and my workplace, which, in several instances, are associated with depression, temporary and long-term disability leaves, and, for two individuals, with complete employment stoppage. I find out that, contrary to what my physician has said, there are acknowledged standard medical clinics specializing in the two illnesses; I find out that, among the alternative approaches, the Vancouver Island Compassion Society sells marihuana to several customers with prescriptions afflicted with one or the other illness; and I find out about a little-studied praxis—sauna—and the benefits to people with chronic fatigue syndrome or fibromyalgia.

Each time I consider my illness, I also think about confirmation bias and hypochondriacism. (Perhaps this is part of my scientific training, but such an attitude certainly does not presuppose my level of training in the sciences or any science at all.) I continually seek a critical stance with respect to myself (a thinking body) all the while attempting to keep my entire Self as the object and subject of inquiry. I want to do research that is not just about me but also for me, providing me with resources that I can use to improve my quality of life. It turns out that doing such studies simultaneously is for and about oneself, such as when a designed change in diet brings about a positive change in the quality of life, leading to both the sought-for change and knowledge about a possible cause of the preceding illness.

After the physician tells me about his hypothesis—he never actually makes a declarative statement such as "This is fibromyalgia" but rather says, "There is a lot pointing to it"—I begin sharing the tentative diagnosis with others. In the course of doing so, I rapidly find out about three other individuals in my workplace diagnosed with one or the other, and one individual at Sylvie's workplace. But there are more, as I find out in the subsequent months and years. In each case, the illness is debilitating to the extent that they take time off work. There are several differences between their situation and mine, which the physician partially relates to my ability to cope better with the chronic pain: (a) I continue to exercise and bicycle to work (the physician comments, "in better shape than 95% of men your age"); (b) being able to work from home, I am in a position to manage bouts of fatigue by lying down, which allows me to keep up my scholarly productivity; and (c) I am able to maintain a positive orientation to life, in part by continuing to garden and cook all the meals, allowing me to sustain the "Zen spirit" that I have developed in the course of my life. This orientation is accompanied by a theoretical attitude—e.g., the embodied and situated nature of cognition—that allows me to understand what is happening even though I am not in a situation to control it. Moreover, the phenomenological interest in attempting to understand such experiences, and the ways in which these mediate what and how we know and learn generally, and in science and mathematics more specifically, appear to be crucial aspects of my coping with the extended and chronic illness.

In the course of the illness, I engage in "experimentation" in attempts to understand whether some aspect of my life (e.g., diet) may be at the origin of a particular problem. For example, for the recurrent problem of a belly distended by excessive gas production, I eliminate a food (e.g., milk and cheese) for a two-week period and then begin eating it again. It is a classic design of $N = 1$ (single case studies), where some treatment is turned on and off again after some baseline has been established. In this situation, my body is not translated into something else, but its reactions are observed in its normal settings as changes are introduced to the normal dietary regime. That is, in contrast to the approaches in both regular and alternative medicine, changes are produced in the individual-in-setting system and any (potential) changes are observed in the same system. These experiments have negative results in much the same way as the tests of the regular doctors. For example, the bloating could have been the result of lactose intolerance. But the elimination of all milk products from my diet—which is easy for me because I cook everything from scratch and I do not eat out—does not bring about a change: the bloating has a different cause. The negative result, however, does not exclude that, in a more comprehensive picture, milk products may be part of the problem. But one particular experiment turns out to assist me a lot with the pain: I grow some marihuana plants and use the dried leaves.

As in so many other cases, I first go to the Internet to see what kind of information I can find; and, as always, I do not stop after one page but seek to find as many as I can on the same topic. I go to government and alternative websites. I learn about the chemical structure of THC (Figure 9.4), short for delta-9-tetrahydrocannabinol, the active ingredient in marihuana. THC was first isolated in 1965 from hemp, known for its intoxicating properties for thousands of years. In Canada, it is grown commercially for medical and medicinal purposes but also, as dronabinol, has made it into appetite stimulants and anti-nausea/vomiting agents used, for example, with individuals undergoing chemotherapy or individuals suffering from AIDS-related anorexia. Using the search terms "cannabis" and "fibromyalgia" in Google turns up many sites with testimonials to the beneficial effects of the drug when everything else doctors prescribe does not help.

I also peruse the research literature by conducting a search in the Thompson ISI Web of Science. There are a number of reports that point to the risk of developing psychosis. But, as in the medical literature generally, the studies reveal little about the actual lives of actual people, and how drug use mediates their lives and decision-making in the particulars of the contexts in which they live. I still remember thinking at the time that the problem with the medical literature generally is that the causal models it produces are based on statistical relations rather than on true causation. Thus, with a sample of sufficient size, reliable effects can be detected that are so minor as to be irrelevant to most individuals. Moreover, knowing that there is a statistically reliable effect does not mean that any particular individual exhibits the pattern. Furthermore, although the general risk of developing psychosis increases, the medical literature tends to overestimate this in the very statistics they use. For example, a recent meta-analysis reports an odds ratio of 1.41 for the development of psychotic behavior in individuals who ever smoked marihuana. Odds ratios, however, tend to over-express effect sizes. For example, if 2 out of 100 individuals taking a drug were to develop psychosis compared to 1 out of 100

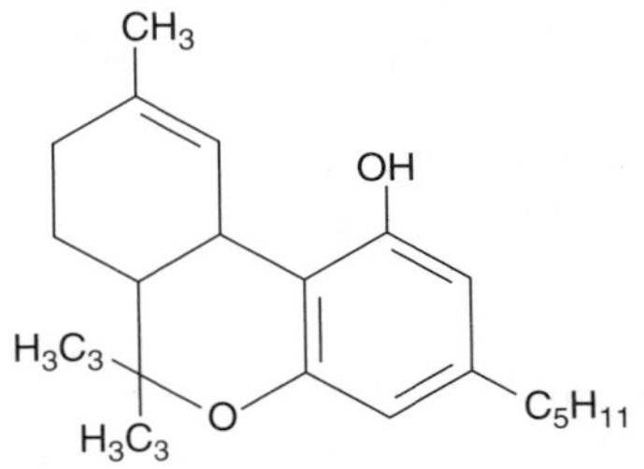

Figure 9.4
The structure of *delta-9-terahydrocannabibol,* the active ingredient that can be isolated from the leaves and buds of the marihuana plant.

not taking the drug, the odds ratio would be 2.02 (according to the equation OR = (2/98)/(1/99)). Despite the ominous looking odds, which double, the actual incidence of the disease remains small. But even more importantly, for any individual, odds do not mean that the person develops the condition. Here again, the translations of real-life experiences into formalisms and the formal knowledge related to them have little bearing on the real life of people. Despite the known risks, I therefore decide to try at least to deal with the chronic pain.

Even though I do not have the background knowledge to distinguish the quality of information on growing marihuana that appears on different sites, it helps me to look for recurrences in the information and for differences. In the present case, I learn how to grow the "weed" in the natural conditions of my garden, beginning with the raising the plants from seeds in my tiny greenhouse and then transplanting them to the plastic tent that they share with some 36 tomato plants. With careful tending and regular pruning and watering, the plants grow to the size of small trees.

One question that imposes itself on me—I do not decide on them, as they emerge within me without my intention—is whether to smoke or ingest the drug. Again, I enact an experiment of the single case study type. Initially, I try to smoke some in a joint. Perhaps because I am not a tobacco smoker, perhaps because of the drug, my stomach turns. (Despite their known adverse health effects, neither tobacco nor alcohol is prohibited.) Furthermore, although the pain disappears, the hallucinogenic effect is so strong that I cannot do anything else but lie down. I decide to prepare some for an experiment of ingesting the drug. I grind up a weighed amount of the leaves and add them to the cookie dough, weighing the exact amount I use to be able to control the effects. There is little effect even with relatively large quantities.

From the Internet I then learn that THC is liposoluble and that the drug acts better when, prior to cooking, it is dissolved in butter liquefied in a bain-marie. Through continued single-case experimentation with the first batch, I find out that ingesting the drug does not have the same effect as smoking it. Depending on the amount, I sense the effect on the body alone—the pain disappears—moving up to the ears, which begin to tingle, with hallucinogenic effects if the amount is very large. I find out the right amount of cookies to eat (the amount of ground up marihuana to ingest) so that the pain disappears without affecting my capacities to write and think. I also find out that, when taken in the evening, I am already asleep when the apparent effects begin two hours later, which allows me to maintain deep sleep, without waking, thereby getting better rest. Being well rested permits me to manage the day without requiring rest, thereby increasing my levels of coping with the illness.

As with my other "single-case studies," there has been little overlap between the formal scientific knowledge I find and understand—the structure of THC—and my everyday life condition, which has been improved. There is little relation between the description of the statistically reliable adverse effects, as represented in the odds ratio for developing psychosis, and the increased control over and improvement of my condition. I have combined careful control of the antecedents (amount of the drug ingested, timing) with careful observations on the effect it has on me; and I alternate from moments of taking the drug and baseline observations. Thus, I exert a high level of control over the experiment, which situates itself in the middle of my life, and therefore immediately embodies applied results. This concrete realization of science from the people immediately has been for the people. Rather than allowing "psychosis" to become a concept that puts scientists and the scientific establishment in charge of my life, I have made conscious decisions about how to find out more when medical practice has abandoned me.

On my next visit to the physician, I tell my doctor about what I have done. He expresses concern with the possible legal ramification. But he does take note (literally, metaphorically) and acknowledges the fact that this is the first time that something has helped me to cope with constant pain and fatigue. He acknowledges the fact that I have been able to concentrate and do more work than before.

Another significant change has occurred after purchasing and installing a far-infrared sauna. As frequently, my research begins after I come across a note that far-infrared saunas have helped individuals with one of the two illnesses. On the Internet I find anecdotal evidence, and there is similar evidence in one of the alternative magazines we pick up monthly at our grocery store. At two local country fairs, I take the opportunity to sit in this kind of sauna and to pick up more information in addition to talking to the merchants.

To me, all of this information is insufficient to buy one of these devices, which range in price upward of $2,800 for a two-seater. As frequently, I use my university's subscription to Thompson ISI Web of Science and search for literature using "'far infrared sauna' AND 'chronic fatigue syndrome' OR 'fibromyalgia'" as my Boolean search parameters. I find several studies—in journals such as *Clinical Journal of Pain* and *Journal of Psychosomatic Research*—that support claims about the benefits of sauna treatment in both illnesses. Encouraged by these findings, I begin an extended search. My wife and I take buying a sauna to be a lifestyle decision. As always, I engage in extended inquiry to find out the differences between models, costs, benefits, service, and so on. We visit a store to see additional models, and eventually decide on buying a three-person model. Within 10 days of doing sauna one hour per day, my muscle and joint pains disappear and I no longer need to seek recourse to the THC-laced cookies. Since then, my health has considerably improved, and though I am not "back to normal," the quality of my everyday life has improved tremendously. It has improved not, as I believe, primarily through any help I might have received from both regular and alternative medicine but through a continued search for understanding and experimenting.

Chronic Illness and Scientific Literacy

After two bicycle accidents in 2001, I have found myself physically and cognitively impaired, without initially linking the accidents and my state. There followed years of testing for different kinds of possible illnesses, including amyotrophic lateral sclerosis. In all of this, I have found myself at the mercy of two medical systems that did not and perhaps could not help me, leaving it up to me to live through a frequently debilitating condition that ultimately received a name: chronic fatigue syndrome/fibromyalgia. In this chapter, which constitutes an institutional ethnographic lens on the medical sciences and medical praxis—I use this experience and my search for a scientific understanding of what was happening to me and of the solutions I envisioned and enacted, which ultimately were associated with a radical improvement of the condition. The experience and my understanding thereby constitute the ground for understanding science and scientific literacy in the everyday life of a person, who is not only led to coping with the situation but also evolving ways of mobilizing science and scientific research reports to provoke a return to wellness.

At what point has my advanced graduate work in the sciences enabled me to cope with my illness and with accessing resources required in understanding what is happening and in engaging proactively in betterment? My tentative claim is that it has enabled me very little. Some science education colleagues might be tempted to claim that my scientific training prepared me in developing a critical stance to information found on the Internet and to studies that appear in the medical literature. I suggest, however, that little of my physics and physical chemistry training prepared me to read the medical literature. In fact, with each article, I had to engage in extended background research just to learn about many of the specialized terms that the authors use to report their findings. At the same time, I suggest that a general literacy focusing on how to read texts critically and how to deconstruct the various forms of inscriptions is a better preparation than the one that students receive in science classes. A general literacy would be able to enact careful single-case studies such as those that I conducted with myself: I easily envision teaching students to evaluate any form of evidence any writer provides for claims and assertions. In part, my assertion is based on the fact that in studying science, students become entrained (perhaps even subjected to ideology and brainwashing) into an ideology of the separation of variables rather than in a holistic investigation of problems, which includes anecdotal and qualitative historical understandings that not normally make it into scientific evaluations. There is precedence for different approaches of handling diverse forms of knowledge, such as including prior knowledge in statistics, as this occurs in the use of Bayesian statistics; and there are approaches, for example, in the reconstruction of marine environments that integrate scientific and qualitative traditional ecological knowledge. That is, in the course of my illness, I have come to learn a lot about standard and alternative medical practice and their ways of accounting for evidence. In my case, some evidence has been excluded systematically, such as the post-exercise ammonia smell and the bloated belly; other evidence has been investigated only insofar as the causes would be located within my body. I also learn about how environmental aspects and life conditions are excluded almost systematically and the problems are sought as being located within the individual, who, as in the case of the naturopathic doctor, prescribes a particular remedy.

In the end, I both learn to cope with the illness and, through careful experimentation and literature research on the Internet and in journals of regular and alternative medicine, eventually find ways that dramatically improve my condition. In saying this, I do not endorse unconstrained growing of marihuana in the same way that I do not endorse unfettered production and consumption of alcohol or tobacco. But it turns out that marihuana helps me at a moment when the regular and alternative medical systems fail to provide me with relief.

To set this conclusion further in perspective, readers need to consider that there is only a small percentage of the general population with the amount of scientific training that I have had. We therefore also need to reflect on the possible mediating effects my advanced knowledge of and about (medical, natural) science has on the processes and products of my inquiries and experimentations. Throughout the process in which the regular medical system has attempted to identify the causes of my illness, I have been both a willing subject/object in the doctors' hands and a keen (ethnographic) observer, finding a considerable degree of similarity in the claims of social studies of science (medicine) and what I see. I have seen and experienced the fallibility of the medical sciences and medical practice in and with my own body and developed a critical medical scientific literacy in the process. More important than getting a scientific literacy in school—what would its content be? What would be the competencies?—is, from my standpoint, the development of an orientation toward continual learning and a willingness to cope in adverse situations. What a curriculum fostering such orientation and willingness exactly looks like has yet to be worked out, but my hunch is that students participating in solving certain societal problems—the seventh-grade students I taught who participated in environmentalism in their community (Roth & Barton, 2004)—develop these attributes. They also develop an ethico-moral stance that currently is not fostered by formal science education.

All the doctors I have seen have gone through extensive scientific training, as medical studies generally require a bachelors of science or three years of undergraduate studies of science in a program often denoted as "pre-med." On the part of these doctors, I also want to see a new form of scientific literacy, which they can appropriate and evolve as part of their secondary and tertiary education trajectories. I want to see doctors routinely take into account and formalize the anecdotal evidence patients do and can provide, enroll them in the recording of life-relevant data, that are to be used in making diagnoses in more holistic ways. Illness, for example, is something that af-fects real people, that is, literally is done (Lat. facēre, to do, make) to (Lat. ad-) someone who is bearing the burden of this doing that he has not wanted. (Chronical) illness demands compassion and an integrated approach that involves the patient and his knowledgeabilities as much as it demands the knowledgeabilities of the physicians and specialists charged with assisting him. I am thinking of the already-mentioned approach to reconstructing the ecological systems in my province, where scientists use quantitative scientific and aboriginal knowledge to make their assessments. Ultimately, doctors and patients need to be involved in more collective efforts of making sense, especially when, as in school situations, the solutions are intended to be more complex than simple solutions that mask the real underlying problems. Much like in science education efforts where all stakeholders come together in cogenerative dialogue situations not only to make sense but also to establish viable change plans, I would like to see doctors develop forms of (social, scientific) literacy that allow them to engage with their patients—whatever their educational and experiential backgrounds—to develop solutions that go beyond simple fixes (iron supplements, vitamin B12 pills or shots rather than more complete articulations of dietary, lifestyle, etc. plans).

Including patients in the production and interpretation of data cracks open the current power/knowledge boundary between doctors (science) and patients (users), because the focus of cogenerative dialogues is the collective production of knowledge. When knowledge is articulated collectively, power comes to be enacted collectively as well. Collective activity leads to the expansion of action possibilities and learning. A collective approach, drawing on the inherently heterogeneous distribution of knowledge within and across participants, expands our possibilities in making (medical) science relevant to the everyday lives of people, especially to those who, as I am, are afflicted with an extended chronic illness. My hypothesis is that a more general critical literacy concerned with reading, understanding, and deconstructing texts of all sorts would prepare me better than knowing a few specific theories and facts (Newton's third law of motion, balancing chemical equations, predator–prey relationships and models). Whereas knowing how to balance chemical equations may never be relevant to my life, competence to read critically is relevant to all aspects of life, which never puts up labels to indicate the specific sets of formal (school) knowledge to apply in the present context.

In the course of attempting to understand the various approaches chosen within traditional and alternative medicine, I have had to understand very different discourses drawing on different sets of facts and concepts. It is virtually impossible both to learn all scientific knowledge presently available and to select any portion of the knowledge that I really required for inclusion in a science or science-technology-society curriculum. Some readers may argue that my science background prepared me to read the scientific literature critically. But to truly read an article critically, one has to be part of the domain and understand which claims are more stable than others. I know about the pernicious effects of pre-figured concepts that permeate and frame our everyday lives through reading books such as *Conceptual Practices of Power* (Smith, 1990) and about researching these effects from *Institutional Ethnography* (Smith, 2005) rather than through anything that I have learned in the sciences at the undergraduate and graduate levels in college and at the university. It is true that in some instances, my physics background has allowed me to assess the claims of naturopathic doctors rapidly, which has led me to instantly take what they said with a grain of salt. But as my extended experience with Internet searches shows, there are numerous ways in which my critical literacy has served me to a much greater extent than any scientific knowledge I have had before and I have developed since.

Notes

1 Although there are multivariate models in the sciences, students are taught (indoctrinated) to change one variable at a time, one of the conceptual blind spots of the standard sciences. This is precisely the point a recent article in the flagship journal *Science* makes, which argues that systems need to be understood holistically (Youngstaedt, 2008).

2 I found out much later than writing the chapter that engineers to do this too, adding their degrees on the by-line in academic journals, something I discovered when publishing an article in an engineering journal on the struggle of local residents to get a connection to the water main.

References

Berg, M. (1997). Of forms, containers, and the electronic medical record: Some tools for a sociology of the formal. *Science, Technology, & Human Values*, 22, 403–433.

Center for Disease Control (CDC). (2006, May 9). Chronic fatigue syndrome. Retrieved August 6, 2008 from http://www.cdc.gov/cfs/cfsbasicfacts.htm.

Latour, B. (1987). *Science in action: How to follow scientists and engineers through society*. Milton Keynes: Open University Press.

Marion, J.-L. (2002). *Being given: Toward a phenomenology of givenness*. Stanford, CA: Stanford University Press.

Mol, A., & Law, J. (1994). Regions, networks and fluids: Anemia and social topology. *Social Studies of Science*, 24, 641–671.

Ricœur, P. (1990). *Soi-même comme un autre* [Oneself as another]. Paris: Seuil.

Roth, W.-M. (2006). *Learning science: A singular plural perspective*. Rotterdam: Sense.

Roth, W.-M., & Barton, A. C. (2004). *Rethinking scientific literacy*. New York: Routledge.

Smith, D. E. (1990). *Conceptual practices of power: A feminist sociology of knowledge*. Toronto: University of Toronto Press.

Smith, D. E. (1999). *Writing the social: Critique, theory, and investigations*. Toronto: University of Toronto Press.

Smith, D. E. (2005). *Institutional ethnography: A sociology for people*. Lanham, MD: Altamira.

Winner, L. (1980). Do artifacts have politics? *Daedalus*, 109, 121–136.

Youngstaedt, E. (2008). All that makes fungus gardens grow. *Science*, 320, 1006–1007.

10 A Stranger in a "Real" Land

Engineering Expertise on an Engineering Campus

Karen L. Tonso

> Your body changes, but you don't change at all. And, that, of course, causes great confusion. (Doris Lessing, speaking about aging)[1]

In this book's quest to talk about science for real, everyday people, there is a deeper issue that concerns why alternatives to "school" science are difficult to promote. I have a vested interest in substantive change in how science educators think about science, because I had an engineering career in the petroleum industry and there were a lot of gaps in my education. From 1972 through 1987, I worked in four different kinds of industry positions: as an engineer doing laboratory and field testing, and on-site inspections, in a foundations-engineering consulting company; for a major oil company in a field production office, where I was responsible for the reservoir engineering side of production operations in a segment of the district; for a major oil company research center, working both in a "tertiary" oil recovery group as the field liaison for a large-scale field test of the technology, and in the reservoir simulator development group writing and testing computer-program mathematical models; and for a regional bank in an energy lending unit performing collateral evaluations. These experiences taught me many things: for instance, that there are no "givens" like those I had seen in textbook exercises, that problems never have one answer, that corporate politics and industry customs trump engineering calculations, that those closest to "doing" industry work hold wisdom that is the only real knowledge available to solve tough problems, and that you had better keep on the good side of the pumper (field supervisor) and ask the right questions, because terse is their native language.

Thus, I am eager to promote alternatives to the norm, but question my success doing so to date. Why doesn't cutting-edge research about atypical science have a greater reform impact? I think the reason is suggested by Lessing's quote: We do, in fact, change curricula, seemingly to account for research advances, but the underlying *form* of science which happens to be seen as the "right" science does not change. Such was the case for a reform-minded curriculum at an engineering campus—Public Engineering School (PES) in the U.S. mid-continent—where I performed a large-scale cultural study in the mid-

1990s. In this chapter I examine these circumstances by following different forms of engineering expertise, that is, forms of scientific knowledge germane to engineering. As explained in depth below, two forms of engineering expertise coexisted at PES. One form aligned with an *academic-science* form of life associated with conventional engineering coursework centered on textbook exercises and timed tests, the other connecting a comprehensive set of understandings and practices from conventional courses with other forms of knowledge that were put to use in novel ways to *engineer* a project, a form of expertise that the campus termed *design engineering.* In my experience and according to studies of engineers at work, *design engineering* seemed better suited for careers as "actual" engineers, student engineers' term for practicing engineers. Reform efforts highlighted this second form of expertise, especially appending to the conventional curriculum a design curriculum where students worked in teams to complete projects for industry and government clients. There was, among student team-mates, much contestation about which form of expertise was "real" engineering and which sort of practitioners were "real" engineers.

I argue that there are two value-sets in play at PES, one underpins academic-science engineering and the other design engineering; the former holds sway on campus, whereas the latter ascends in industry settings. With 15 years of industry engineering experience (and depending on ethnographies about engineers at work), I readily recognized *industry* engineering expertise in *design engineering* when I observed student teamwork. It seemed likely that the *academic-science expertise* privileged at PES (and other campuses) via campus traditions, routines for success and excellence, teaching practices (especially learning activities and grading), and other curricular and cultural norms would not be the sort of expertise desired by "old-salts" responsible for moving young engineers into careers quickly with minimal expense for "learning from mistakes." In spite of industry concerns about the preparation of engineers that suggest rather pointedly that campus preferences for academic-science engineering expertise promote a kind of engineer who will not "hit the ground running," but will require one to three years of on-the-job training, conversations about which form better prepares students for the work of engineers have not been convincing to many engineering faculty. In fact, new "design" courses failed to shift what counted as engineering expertise in the PES culture, and this is the first part of the tale I tell here.

The second part of the tale involves how difficult it is to argue that this alternative form of engineering expertise *should* be given more credence. If industry work is taken to be real engineering, the engineering preferred on campus will be strange indeed. Thus, this chapter is also about how one convinces an *academic* faculty like that at PES that they elevate an engineer who will be a "stranger" when s/he enters industry careers, or figuratively "real" land. Ultimately, the argument about what is "real" and what not influences how my argument (and that of other engineering colleagues calling for reform in engineering education to better prepare engineers) is taken up by many engineering education insiders. Ultimately, the second part of the tale concerns the complexity of making a counter-claim when those hearing the argument are using a taken-for-granted value-set—without recognizing that they are doing so—and recasting my argument into an image of the one I intended to critique. (Shades of dismantling the master's house with the master's tools!)

Public Engineering School

Public Engineering School is a state-supported, co-educational college with programs typical of those at many engineering colleges. At the time of the study in the mid-1990s, undergraduate engineering enrollment was close to 2,300 students, about 14% ethnic minorities. Women students comprised over 20% of the undergraduate enrollment, somewhat higher than the 18% average nationally. The campus was chosen because it stood out as an engineering college, with more women students and professors than national averages, as well as because it possessed considerable collective will to address concerns about women's education in engineering. About a decade before my study, in the 1980s, the campus promoted learning more about the "practical" side of engineering by adding engineering design classes to their curriculum. Every student took design courses during their first, second, and fourth years of study—a total of six semesters of study, about 11% of their total course load. By attending design classes and working as an engineering colleague on teams at the first-, second-, and senior levels, I came to know 33 students (15 women and 18 men) and 11 professors well. A rich sense of campus culture emerged from primarily ethnographic research strategies, which provided a way to study social interactions during teamwork sessions and class time, how those interactions fit into and were influenced by classroom (and campus) activities and practices, how students thought about design courses versus non-design (conventional) courses, how students talked about and organized ways to belong as an engineer on campus and how to do engineering, and the influence on students' day-to-day life of campus curricular structures (the interconnections among individual courses into a trajectory leading toward graduation) (reported as a cohesive whole in Tonso, 2007).

As with most institutions, students, faculty, and administration at PES used a rhetoric to describe an imagined reality, or normality, that was a significant over-simplification of everyday life on campus. This rhetoric encoded the right kind of engineer and engineering, and was the standard against which to measure other ways of being and forms of expertise. However, throughout what follows, I trouble these waters using not only my deep sense of what engineers do in a very wide range of industry employment settings, but also an expansive ethnographic literature about how engineers do engineering in real-world settings after they graduate; that is, I suggest that there were two value systems at play on campus. The first was "school science" and it is the one that held sway at PES; that is, it was hegemonic locally. Industry engineering provided a second value set and I argue that this form of expertise and way of being an engineer held sway in PES design class rhetoric, at least, and *should* play a greater part in science education.

Expertise

Expertise, here, is taken to imply the forms of knowledge needed to *do* engineering. Rather than understanding such a concept as static and universal, I suggest that there is considerable variety not only in how engineers go about doing engineering, but that different forms of expertise are preferred in different contexts, and that this has implications for what school science *should* be, if the goal of school science is to prepare students for the meaningful use of scientific knowledge. Here, I begin with a brief recap of engineering expertise drawn from ethnographic studies of engineers at work, then move on to consider forms of expertise at PES.

The Work of Practicing Engineers

Fine-grained, ethnographic studies of engineering work across a wide range of employment settings suggest that, though academic-science concepts prove necessary for engineering, they will take a back seat to other forms of knowledge and skills. The complexity of engineering activity was more expansive than conventional curricula envisioned, and, contra a tendency to parse the technical from the social, it became clear that engineering was simultaneously technical and social. "[Su]ccessful technological innovation involves the construction of durable links tying together humans and nonhuman entities ('actors')" (MacKenzie, 1996, p. 13). "Real" engineering is a profoundly complex set of beliefs, knowledge sets, skills, and practices that differ from the universal, generalizable knowledge of academic science. Engineering expertise surely has some aspects that depend on academic-science, but when engineers deploy these understandings they do so in ways that are contingent and contextual. And, the term "heterogenous engineers" (coined by Law, 1987) captures this reality, that technological work, of which engineering is a subset, requires practitioners to deploy not only technical expertise, but also other forms of knowledge of a more social nature (e.g., outreach, marketing, and fund-raising efforts, for instance, a list that has only grown as researchers continue to study engineers in a wide range of employment settings). Based on studies of engineers at work, those PES students with design-engineering expertise seem to be better matched to engineering work.

Engineering Expertise at PES

At PES, engineering expertise came in myriad variations with two distinct end-points: *academic-science expertise* aligned with conventional engineering coursework and *design-engineering expertise* intended to be developed during design courses. The academic-science form had prominence on campus and, even in design courses, the design form was little recognized by cultural practices. Though students began to cobble together their own expertise as soon as they began their studies, it was only at the senior level that striking differences in capabilities became apparent in the extreme cases. That is, some students were remarkably adept at those skills needed to correctly answer all problems on a test; others might be so-so in their conventional coursework, but excelled at deploying this kind of information, along with what they learned from plant operators and on-site engineers about business practices (for instance) in a real-world project; while others were equally as adept with academic-science knowledge as they were putting together a wide range of other forms of knowledge needed to understand a particular real-world project and successfully complete it. Thus rather than these forms of expertise being disjoint, the very best design-engineering practitioners also had deep understandings of academic-science knowledge, understandings in fact that easily exceeded those of their team-mates who had higher grade-point averages and more campus status.

Academic-Science Engineering

Above all else what distinguished prototypic academic-science engineering expertise was the narrowness of the knowledge set with which these student engineers worked. Here, "expertise" meant solely understanding the formal knowledge made evident in conventional courses, where textbook exercises had one right answer that was derived from "given" information, and where tests incorporated a suitably difficult range of such exercises. To have academic-science engineering expertise, then, meant being able to manipulate equations, to read unknowns from problem statements, to apprehend the world in its most stripped down form. At some point in an engineer's education, being able to connect these textbook exercises to the world, to anticipate which over-simplifications must now be accounted for, to measure key facts needed in engineering equations, and to select the right equation for the right situation becomes more important than getting the right answer as quickly as possible. These are skills that an engineer must have, but that a student displaying academic-science expertise lacks. For instance, teams in the first-year course designed a gate that could be opened only with a vehicle. Two of the three teams developed a gate that their vehicle drove over, and on both teams, older students (with considerably more construction work than their just-out-of-high-school team-mates) suggested this gate (or an early version of it). However, none of the students made sense of (computed) the forces available to open the gate, in spite of the fact that all had seen this idea in abstract form in earlier courses—the sliding block (car) on an inclined plane (hill), and the design professor failed to make this connection.

Furthermore, academic-science practitioners could rarely deploy their academic-science knowledge set in their design work. Consider the cases of both Russell and Jessica, who had high grade-point averages, indicating that they were academically successful in conventional courses, but whose design-class work did not live up to their reputations. Contrast Russell's actions with those of Martin and Marianne on the day we visited the sludge disposal system at the power plant:

> Martin and Marianne scrambled over the piping to work out the flow processes and follow electrical conduit from the pumps to meters to the controller box. Russell looked confused. He asked me to explain what we were seeing. (What do I know about a power plant?) At first I thought that he was toadying up to me, but he really did not have any idea what the pipes, pumps, electrical conduit, meters, and gauges were all about.

On another occasion, a team-mate suggested using an approach exactly like something covered in a required course, and ostensibly something Russell was able to accomplish at one point in time, since he reported getting a high grade on that lab. But, he could not follow the discussion, requiring that Martin provide an explanation that not only took Russell back in time to the lab course, but also reminded him of the particular lab activity, and then detailed the academic-science concepts being taught on that particular day and how they related to what was going on in the plant.

Jessica, an electrical engineer, had similar disconnects. For instance, when the team began estimating parameters needed for their installation, Jessica volunteered to calculate the resistance in the electrical wires that carried a pressure transducer signal from the basement to the control room, a precursor to selecting a voltage source for their installation. To make this calculation, she needed to know the resistance factor for the kind of wire that would be used. Her first resistance estimate was extremely high and did not make sense to her or her team-mates. She sought the advice of an electrical engineering professor, who pointed out that she was using the resistance factor for coaxial cable, instead of using the lower value for copper wires. This was a fairly routine calculation and something that all electrical engineers, and most engineers from other specialties, should be able to perform. Like other students who could demonstrate academic-science expertise in their courses, she and Russell failed to carry them into their design projects, to connect these understandings to the real world during design work.

Thus, students practicing academic-science expertise demonstrated formal knowledge primarily in the context of their conventional courses and seemed especially ill-equipped to do so in the context of their design work.

Design-Engineering Expertise

Students who demonstrated design-engineering expertise linked academic-science understandings learned in conventional courses with a deep appreciation of the place of this formal knowledge in the world of design projects. Martin and Marianne, last seen scrambling over the sludge-disposal system, exemplified those with such skill. Listen as Marianne explains her interpretation of a conversation between their client and a faculty member to team-mates:

> I think they're talking about two different things. The client [Curtis] is talking about new technologies. They [the manufacturing company] invested money to maintain this old equipment and then, when the new technology comes along they've got this perfect equipment that's completely outdated. So the cost [to upgrade] is that much more; they have to buy the new technology, plus they've spent all this money [on maintenance], keeping up the cost. Dr. Edison is talking about PCs [personal computers], and sensors, and predicting when it [the equipment being monitored] will fail.
>
> For the work we're doing with hydraulics [in the lab where I'm one of the mechanical engineering professor's assistants], we have pressure sensors [that we monitor]. If you get a line that's crimped, that's going to cavitate a pump [a damaging condition], the PC [uses the data being acquired and] can shut the pump down. Then you go in and fix [the line that's crimped]. So Dr. Edison is talking about a situation where sensors know when to shut [equipment] down . . .

Thus, in the first month of their work at Private Power, and long before the team visited the sludge disposal system, Marianne could already appreciate the company's dilemma and frame it in terms of a faculty member's research agenda, something that Jessica and Russell could not do in spite of industry internships the previous summer. She therefore deployed what she knew from campus in a lab environment to the power plant site for this team's project. Here, she was working out an understanding of the client's business practices, a knowledge set encompassing knowledge associated with a particular context (or contingent knowledge) and knowledge gained from what an insider to the situation knows (tacit knowledge). She combined different forms of knowledge that went beyond formal knowledge learned in conventional courses. And, in the overall scheme of design-course grading strategies, these understandings will count for naught.

Martin, too, possessed the capability to make these kinds of connections. When thinking about the kind of equipment they might need for their project in the second month of their project, Martin ranged across several forms of knowledge.

> We would plot power versus time. (He draws a Cartesian plot with time going to the right on the x-axis, and energy consumption going up. He sketches a wiggly line going to the right.)[2] It's not the same with vibrations. On vibrations, you're looking at frequency versus time, so it's more important that we look for a frequency spectrum. With an accelerometer [a device for detecting vibrations], you get this time-domain data and you have to do a Fourier transformation [a mathematical technique for

> characterizing certain kinds of complex data by using simpler, usually trigonometric, features to get at components of the vibration spectrum] to transform that time domain data into frequencies.
>
> So we get, for instance (and then he draws another Cartesian plot and this time he draws a sine curve up above, in the first quadrant above the x-axis, gently going up and down and very regular, and he continues), if you do this then you get one spike per sine wave. (He draws another plot below it that I cannot see.) . . . If the data's real [expletive for messed] up [and] it's got a lot of warbles in it, then you're going to get components of tiny warbles. (And now he draws a very squiggly line following the sine curve. I imagine it as looking like a sine wave with a perm.) So we need an accelerometer.

Martin is an engineer speaking to knowledgeable colleagues about the fine points of focusing their project on preventive-maintenance, and doing so by linking academic-science (or formal) knowledge, along with what he knows about equipment (instrumentalities, both things that vibrate and instruments to detect vibrations) to their particular project—that is, he deploys design-engineering expertise by re-contextualizing decontextualized knowledge learned in conventional courses and to do so requires several forms of knowledge (Table 10.1).

In addition to Martin's linking formal knowledge of Fourier transforms to instrumental knowledge of pumps and accelerometers, he used informal knowledge about vibrations and connected these understandings to the way that Cartesian plots can be used to represent them, another expression of formal knowledge. Marianne's explanation was grounded in tacit knowledge learned in her work as a research assistant for a professor, as well as contingent understandings connected to the context of their project. In addition, her ability to

Table 10.1. Components and Contexts of Knowledge

Components of knowledge	*Embodied in . . .*
Formal knowledge: theories, formula, available in written form, e.g., textbooks, acquired through formal education	Codified theories
Instrumentalities: knowledge embodied in the use of tools and instruments, learned through demonstration and practice	Tool use
Informal knowledge: rules of thumb, tricks of the trade, sometimes available in guidebooks, etc.	Verbal interaction
Contingent knowledge: distributed apparently trivial knowledge, acquired by on-the-spot learning	The specific context
Tacit knowledge: rooted in practice and experience, transmitted by apprenticeship and training	People
Meta-knowledge: general cultural and philosophical assumptions, acquired through socialization	The organization

tack back and forth between what a professor and a client contact said demonstrated her skill in deploying understandings about the ways that organizations interact, meta-knowledge of interactions between campus research activities and power plant maintenance expenses.

Thus, rather than academic-science expertise and design-engineering expertise being two disjoint arenas of technical understandings, *academic-science understandings are integral part of design-engineering expertise.* Taken together Martin's and Marianne's explanations connected academic-science understandings with the real-world circumstances of their project at Private Power. In fact, they ranged across all of the knowledge domains deemed necessary for practicing engineering (Table 10.1). And, notably, students who learned how to gather and deploy understandings from all of the forms of knowledge not represented in their conventional classes learned important social-interaction skills that will prove central to their engineering work, and that will facilitate continuing to learn from other people, in other engineering sites and business or social contexts, and about other kinds of equipment.

Consequently, deploying design-engineering expertise is a more complex endeavor than the design-course rhetoric of "applying formal knowledge to a real-world problem" seemed to imply. First, expertise requires continuously expanding knowledge sets (what I suspect is the sort of lifelong learning required of an engineer), especially linking what was learned in conventional courses with what an engineer learns about the real-world setting where the dilemma being addressed resides, along with what an engineer might bring along from other places—the former construction workers' drive-over gates in the first-year course; Marianne's experiences in her assistantship; and Martin's understandings of the relationships between Fourier transforms, vibrations, and pump maintenance—those similar experiences that suggest ways to proceed on a new project. Second, design-engineering expertise is necessarily local or contextual, and constructed through social interactions among people in the site and other experts or advisors. The social-interaction dimensions of design-engineering expertise—deploying skills not only outside the intended curriculum at PES and only minimally supported in design courses—proved central to demonstrating expertise. That is, students discussed the power plant operation with control-room operators, the company engineer, and the plant manager, as well as worked through complicated engineering concepts and applications with faculty advisors.

As such design-engineering expertise is a rather unruly beast, and its monitoring and recognition seemed to lie outside conventional PES engineering education practices, especially what is usually termed "grading." In fact, at PES, the engineering work of students by and large did not come under the scrutiny of faculty, who focused instead on superficial aspects of written products, checking for instance: if all of the required sections were present and in the right order, were formatted as directed, were written mechanics clean enough (spelling, grammar, punctuation, and the like), and in the final analysis (if one watched how students accomplished these things) had students produced the image of having done what they were told to do. A few teams "boiler-housed" (made up) results, such as the seniors who, finding that their colleague would be marked down for her section, created a new section out of thin air to mask that their team-mate had done little on her part of the project over the course of the year. One first-year team realized that they had never developed a list of criteria against which to rate their gates and select a best gate, so at the eleventh hour dashed something off, quickly manufactured ratings for each criterion for each gate, typed this information into a table, and provided at least the illusion that their decision had been guided by carefully considering to what extent their gate met the criteria set by project specifications and the team. In fact, their gate would not open, so it was hardly a robust gate design, but this did not preclude their writing skill garnering the first-place award among the teams in the professor's three design sections.

The campus rhetoric to value design therefore had not extended into matching assessment to course activities, or to what students must be able to do to be considered engineers. Here, grading practices created a minimal standard for what counted as knowledge worth having, even when design courses offered very authentic ways for students to deploy their textbook knowledge, along with several other forms of knowledge. These other forms of knowledge were not graded. That is, unconventional skills and knowledge sets came to be devalued via not being given standing as something worth grading. Ultimately, a tension developed between academic-science and design-engineering expertise. And this tension could be routinely seen in the day-to-day interactions among students on design teams; that is, student engineers contested which of these two forms was *real* engineering.

Contesting "Real" Engineering

Contestations about what constituted "real" engineering became fiercer as students moved from first-year to senior students, and were more prominent when students held extreme views on the importance of academic-science engineering.

FIRST-YEAR TEAMS

Two first-year teams took quite different tacks on their project. The Monday Team (three women, two men) simply went through the motions: devoting as little effort to the project as possible, meeting course deadlines, and attending primarily to graded products. They rarely met outside of class and when they did almost no design work occurred, as suggested by this team meeting:

> During one rare meeting, called to prepare for an oral presentation when students seemed genuinely concerned about their ability to carry off the presentation, only three students arrived close to the starting time of 8:30 p.m. While they waited for the meeting to start, they conversed about

> typical student issues. Blayde maligned Dr. Davis' teaching in another course, and Mary and Marika discussed other classes and remarked on Clark's saying he would be there. But only two comments in 30 minutes of time dealt with their design project: "We have three people and four things to do during the day on Friday. I'm not going to go to calc[ulus] [presumably the class where something *wasn't* due]." And: "Here's the picture (and hands her a gate sketch that I cannot see from where I sit)." Angela arrived after 9 p.m. without Clark, but carrying his apologies for being late: "Everybody's having problems copying a form down [from the file-server in the computer center] and Clark's ready to kill." Clark never made it to the meeting, because of unexpected problems in the computer lab. (He hoped to finish in the computer lab in time and was reluctant to relinquish his place at a computer, which would mean returning and having to wait for a computer.) Ultimately, the students agreed that each would cover the same topic as at an earlier presentation and left.

No one on the team questioned that assignments, like the computer lab work, should take precedence over design, so there was little controversy about their approach to design work.

But this was not the case for the Friday Team (two women, two men), whose work together was grounded in a commitment to everyone on the team being able to perform any of the tasks required for succeeding in the course. This was especially important to Kyle, the former construction manager. As a full-ride scholar recognized for his past industry experiences, he knew not only that not doing well would have a negative impact on his future, but also that he could easily perform all of the work in the team, yet would not have time to do so. In addition, he expressed great respect for young engineers and for the importance of their beginning to see how important being able to connect their coursework to the world would be; that is, he advocated design courses.

Kyle immediately set a tone of their time together being serious time and their commitments to the team being serious commitments. The team met every week outside of class; everyone attended; students took turns chairing meetings; they worked from an agenda and ended each meeting with what they had committed to do for the next meeting. This meant that all of the students agreed to arrive, for instance, with gate designs. As they worked through the various tasks required to complete the gate, only Ron disappointed the team by not doing the agreed-upon work. This led to a dust-up in the team between Kyle and Ron, and Ron gave excuses about needing to do work from another class, an argument grounded in his affiliation with academic-science engineering. But this argument held little water on an engineering campus where everyone took the same courses in their first year and had the same obligations. Ron never missed another commitment.

SENIOR TEAMS

At the senior level, the Sludge Team leaned toward design engineering and contestations were low key, but on the Mercury Team a powerful student affiliated with academic-science engineering kept teamwork in turmoil for most of the two-semester course.

On the Sludge Team, Marianne and Martin clearly possessed design expertise, Jessica and Russell demonstrated academic-science expertise, and Nate gave considerable attention to the academic-science aspects of student life, but hoped to become a better design engineer. Nate accomplished this by working closely with Martin on several aspects of their project, thus Nate learned through something of an apprenticeship to Martin, whose expertise with computers was legendary. Queries about what the teams should *really* be doing came from Jessica and Russell. These students who affiliated with academic-science expertise took a back seat performing engineering tasks, and instead involved themselves with non-technical work and faculty directives—asking, for instance, when the next graded product was due, or worrying about whether the team would miss a deadline. Though an unequal work distribution developed, Jessica and Russell did not deny the skill of their design-engineering team-mates, interfere in their work, or claim it as their own when faculty were present. Yet, when the team participated in classroom and other campus activities, they signaled (primarily by inaction in design work) that academic-science expertise held sway and conceded that people like Martin and Marianne were being overlooked and their skills not recognized.

In contrast, the Mercury Team became a place where overt power moves were commonplace. Four members of the team were academic-science practitioners: Pete and Carson became overbearing, and Carol and Shane passive. (Carol's section of the report was the one rewritten by Pam and Samuel to prevent Carol's being graded down.) This left Pam and Samuel, design engineers, performing most of the team's work. Pete and Carson lobbied for minimal work on the team in general. Pete preferred activities that returned rewards solely to him, as illustrated by this field note excerpt from a team meeting that he was supposed to be chairing:

> Pete left the team meeting in the computer center, where he was supposed to be guiding the writing of their co-authored preliminary design report. While his team-mates sat at a cluster of computers typing text to meet a rapidly approaching deadline later in the day, he appeared with a group of prospective students and their parents, giving a campus tour.

In comparison to his team-mates and to other senior students (and through careful analysis of all of his teamwork contributions, oral and written), Pete routinely failed to demonstrate engineering expertise of any sort during teamwork, obfuscated important issues with baseless claims, gave conflicting facts in the same explanation, confused units of measure when trying to state facts, and took almost no responsibility for his part of the project—eventually giving a vacuous explanation of the condenser he ostensibly spent two semesters designing. Ultimately, most of this was outside the purview of their faculty advisor and could not count against Pete. He was marked down only one mark, from A to A-, based on his team-mates' end-of-course peer evaluations, and this tarnished his perfect grade-point average.

When team decisions about how to proceed occurred, Pete and Carson used excuses, such as "changing at this late date would be too much work," to promote continuing on a plan of action which overlooked not only key constraints of the project but also their colleagues' expertise. In time, avoiding change led the team down a path that collided with a project constraint and it became clear that the team must change course. Pete, Carson, Carol, and Shane minimized their involvement in that effort and kept the workload on Pam and Samuel, who had already done most of the work. Pete asserted to their faculty advisor that he played a central role in solving the dilemma, an untruth, since my field notes documented that he could not explain the dilemma or its solution. Pam and Samuel's ability to argue for the importance of design-engineering expertise was impaired by the campus recognition given students like Pete and Carson, and by the fact that one of them was a woman, someone her academic-science practitioner colleagues thought of as "a unique individual," a person they could dismiss, refute, and otherwise undermine, even as they depended on her to do their share of the work. Thus, a thoroughly lopsided logic developed wherein academic-science engineers acted as if academic-science expertise should prevail and design engineering was not a valid form of expertise, while expecting that the design-engineering expertise of their colleagues would carry the project through—and ensure a good grade in the course.

As was the case for the first-year course, neither the varied forms of knowledge nor their complex interconnections when deployed in design projects came to light via course grading practices and this had implications for whose version of "real" counted. Senior students received marks for oral presentations and written products. In oral presentations, arcane rules, such as standing by the screen instead of the projector or following tight time limits, made little sense to students, and vacuous or erroneous explanations fell below faculty's radar, in spite of indications of deficiencies in company engineers' comments. For reports, faculty again focused primarily on instrumental issues: formatting, type and order of content, and writing mechanics. Little was required of the technical or engineering issues, except that they not raise faculty's suspicion or indicate that one student or another was not a "team player."

Over time on campus, therefore, preferences emerged that signaled the ascendance of academic-science expertise over design expertise, and in toe-to-toe contestations about which form was "real," academic-science won out, at least on campus. Several aspects of campus life contributed to this. First, the campus curriculum gave prominence to academic-science activities, especially fast-paced courses and heavy course loads devoted almost exclusively to a canonical problem-solving algorithm that has only limited relevance in many engineering careers (e.g., Bucciarelli, 1994). Second, routines for success and excellence remained firmly grounded in easy-to-measure concepts and skills, and a devotion to independent work, making timed tests the preferred assessment strategy. Third, teaching practices tended to focus on articulating in a fine-grained way what was required in graded products with less attention to what was required of high-quality engineering. And, finally, grades were the ultimate arbiter used to ascertain who was "best." Scholarships, on-campus awards, and interviews with on-site industry and government employers depended on grades. Thus, students like Pete—who demonstrated so little engineering expertise that I thought his team-mates were pulling my leg when they disclosed his high grade-point average, went on almost 15 plant trips (an on-site visit in the hiring process), had almost as many job offers, accepted a lucrative offer in January, and started driving a company car soon after—held real power to make their version of "real" stick. (And, with his grades giving him better access to industry employers seeking to hire graduates in his discipline, Pete reduced opportunities for Pam and Samuel to interview for those same jobs.) But, Martin, whose readiness for engineering work was far superior (from my vantage point), had far fewer on-campus interviews, went on only two or three plant visits, and took an acceptable, but lower-paying, offer late in his senior year.

Talking across Paradigmatic Divides

My paper presentations seem to have been mis-heard. Consider an early paper I gave at a national conference of engineering educators. Using a wide range of evidence from fieldnotes and interview data (from the sophomore design class of the larger study discussed here), I developed an argument that illustrated how student engineers learned to see the world the way it is *supposed* to be (its cultural tenets) and failed to notice that they also saw other things, but did not recognize them as contradicting cultural beliefs.[3] Given in a conference ballroom, there were microphones in the aisles where questioners immediately lined up after my talk. The session chair began to call on each person in turn, and one after another professor of engineering commented, "I've been a professor [for some number of years], and never seen anything like what you describe." Since there was no question to be answered, I thanked the speaker, then after five or six commented in this vein, I asked for a show of hands of those still waiting to make a similar comment and acknowledged them collectively. I was baffled at their presumption that their personal experience could negate my findings. (I wondered if they simply did not understand the kind of access that I had to sites where they have no access, or if they simply missed the point of the paper.) I responded to the gathered group something to the effect that this was precisely my point: being an insider means learning to take for granted the way things are *supposed to be* and I could not tell (without fieldwork, which would be impossible at this point) whether the professors' experiences differed from those of my participants or whether, like my participants, these professors were seeing through a cultural lens and were unable to recognize other interpretations of events. Session attendees seemed unable to let go of personal perceptions about "normality" and this appeared to blind them to the paper's findings.

Recently, when presenting a paper at a session devoted to engineering education research, I experienced a similar reluctance for engineering educators to believe empirical results. I was discussing the set of findings presented here, findings reviewed in a peer review process for acceptance to the session. Reviewers of the paper and session participants expressed concerns along two lines: they denied that design-engineering expertise was more closely matched to real-world work than academic-science engineering, or they argued I had "gone native." That I might have gone native seemed an odd comment deployed simply to bring into question the research method. Ethnography is a research practice in which few attendees had experience and, even though social science researchers have not been worried about this, I spent an undue amount of time countering the "gone native" concern, without notable success, seemingly. Their argument boiled down to the claim that if I had been an engineer and if I was now an ethnographer doing research in engineering education, there was no methodical strategy or set of strategies that could prevent my "going native," my research practices and anthropology's long tradition notwithstanding.

The other concern seemed troublesome because it required readers to dismiss empirical findings from other studies, or so I thought. I had grounded the claim about the suitability of design-engineering expertise for on-the-job engineering first in a large—and growing—literature and second in my personal experience as an engineer. I found it curious that reviewers and session attendees, however, focused almost solely on my experience (ignoring scholarship about the work of engineers). As the discussion continued, tensions seemed to develop along the same lines as those I found in my research—academic-science engineering adherents took for granted their ascendance relative to practitioners affiliated with other forms of practice and, as a way to bolster their argument, challenged my expertise *as an engineer* and highlighted their expertise as *engineering educators.* I would be remiss if I did not mention here that I was in fact beginning to feel very much like an outsider, an interloper if you will, in engineering education and began to suspect that I had hit a rather raw nerve. It seemed I was being interpreted as a threat to engineering education, or at least to some form of it, and was being actively denied a place in discussions about engineering education, a rather familiar sort of circling the wagons I had seen Pam (a senior student) and women faculty subjected to.

These kinds of encounters point up the challenges of developing a critical vantage point on a powerful discipline, and they suggest some of the dilemmas that become apparent when scholars attempt to talk across paradigmatic divides. I am fully aware that many engineering educators do not accept my research, in spite of the fact that I seem to not only make perfect sense to social science scholars studying science and technology, but also to be well thought of in those arenas. How does one make a persuasive argument across a disciplinary, and by implication a paradigmatic, divide between social science researchers working in the post-structuralist/postmodernist moment, and engineering education faculty assuming a positivist stance?

This dilemma seems to come to full flower when an engineering educator *transacts* with my writing, as reading theorists term the social construction that occurs when a reader with certain kinds of experiences and ways of viewing the world engages with a particular text. I have written from an interpretivist stance, one that explicitly challenges epistemological axioms associated with positivism. Here, I am in particular thinking along the lines of Charles Taylor in his watershed essay "Interpretation and the science of man," as summarized by Kenneth Howe (1998):

> [H]e [Taylor] rejects the view that there can be any scientifically neutral, impersonal language (a central tenet of positivism) with which to describe and interpret human activities. Rather, he says, "we have to think of man [sic] as a *self-interpreting* animal . . . [T]here is no such thing as the structure of meanings for him independently of his interpretation of them."
>
> (p. 13, emphases are Howe's)

The world of social science researchers has been engaged in an interpretivist turn since the late 20th century and my work fits into this paradigm, with ways of making judgments about rigor.

Without belaboring the point, academic readers of empirical research must read with ideas about rigor in mind and must make judgments about rigor as a part of the reading. For instance, several sets of criteria for research quality have been suggested for judging rigor, of which I use four to illustrate the point. Among the early offerings were credibility (both activities in and out of the field), applicability, dependability, and confirmability (Lincoln & Guba, 1985). These criteria suggest that quality is designed in, that it is carried through by checking and rechecking with participants to be sure that their perspective and not the researcher's is presented, and that it is enhanced when multiple researchers or peer reviewers can provide some measure of oversight on research activities, data interpretation, and writing products. Extending these four to account for the fact that educational research encompasses a wide range of disciplinary associations and that progress within each differs from the others, other approaches came to light. Some argued for taking these disciplinary issues into account by meeting central guidelines for the research (Howe & Eisenhart, 1990):

- Fit between research questions and data collection and analysis techniques
- Effective application of specific data collection and analysis techniques
- Alertness to and coherence of background assumptions

- Overall warrant (balancing and going beyond the first three, being alert to and able to employ knowledge from outside the perspective being used)
- Value constraints, both its worth for improving education (external) and research ethics (internal).

And finally, a set grew out of studying qualitative research studies themselves (Shank & Villella, 2004):

- Investigative depth (research that goes below the surface and gets at deeper issues)
- Interpretive adequacy (forming complete and complex understandings of what is studied)
- Illuminatory fertility (what are the implications of these findings in a wider educational context?), and
- Participatory accountability (how have researchers interacted with participants and what are the results of such involvement?)

Each of these sets of criteria recognizes that researchers are *in* the field, not separable from it, and thus must engage methodological strategies to reduce researcher bias—to improve the extent to which research findings represent views of research participants and have not been unduly influenced by the researcher. Though interpretivist and postmodernist researchers debate these sets of criteria, there is general agreement about the underlying axioms underpinning them and an awareness of the reasons for eschewing a positivist approach to social science research. That is, there is a common ground upon which to have a debate about findings and upon which to converse about the implications of research studies, to ascertain which ones really matter, or should matter.

What happens if a reader lacks such underpinnings? There are demands on a practical argument (such as those that apply in morals and politics) that "starts off on the basis that my opponent shares at least some of the fundamental dispositions toward good and right which guide me" (Taylor, quoted in Howe, 1998, p. 15). That is, we cannot begin to discuss research quality, for instance, sans a place to begin. Here, then, is where I seem to run aground in my quest for a persuasive argument: an argument is not something that simply emanates from an author, but is something socially constructed between author and reader. Being able to read research results, then, seems to require that writers and readers have an appreciation of a wide range of thought and an openness toward ideas that are new and that may challenge traditional views. Thus, sense-making is a shared responsibility and my part of the bargain is to describe how I have framed and carried out my study and upon what basis I can make claims to know. But, mis-readings or mis-hearings of my research seem to lack this beginning point. Readers seemed to lack an acceptance of the fact that I had developed a kind of "fly on the wall" presence at PES, and that my findings were grounded in data that necessarily lie outside the range of possible experiences for engineering educators.

If some readers' lack of understandings about rigor leaves them to depend on their personal experiences to judge the "rightness" of findings, we are stuck. This may be what happened in the incident of faculty lining up at the microphone. Did engineering education faculty "read" research findings subjectively, gauging findings not based on disciplinary association or research rigor, but rather asking if the findings map onto readers' (listeners') experiences? If so, how can such readers judge findings if they are precluded from having the kinds of experiences that I can have as researcher voyeur? And, if other readers assumed positivism, would that not be tantamount to assuming academic-science engineering *should* be preferred? Thus, if a reader worked from an unexamined stance that "science" as reflected in positivist axioms and in the kinds of ideas, concepts, and practices of conventional engineering courses was reality, then there seems little chance for a fair appreciation of either my research method or of my claim that design engineering should receive more attention at PES.

I wonder if, ultimately, lacking a common grounding in research underpinnings, comprehension across a paradigmatic divide became troublesome, and part of the difficulty grew from the enculturated power structure within which discussions about reform can occur. Not only were research claims *to know* under attack, but also my being a person with standing to contribute to what counted as engineering or engineering education seemed in dispute. In spite of the lack of robustness to critics' counter-argument, in a setting where there were far more seemingly ill-founded comments than there were listeners able to judge rigor adequately, my argument—that there was indeed a very important sort of scientific knowledge (design-engineering expertise) made evident in students' engineering work and it deserved more attention in grading schemes—carried little weight. Thus, even as I argued that Public Engineering School preferred a certain form of engineering expertise that produced certain kinds of engineers who would be "strangers" in "real" engineering work and simultaneously produced practitioners of design-engineering expertise as strangers on campus, it felt as if I was being read as a person who wasn't *really* an engineering educator and didn't *really* have much to say about reform in engineering education; that is, I was being made a "stranger" to the conversation about forms of scientific knowledge, and ultimately to reform in engineering education.

Conclusion

My two-part tale suggests how the power of the scientific stance made manifest in positivist axioms holds sway against empirical evidence to the contrary. That is, though the notion of universal, generalizable Truth, writ large, is but a social construction resulting from a particular set of agreed-upon conventions (positivist axioms), there is power built into science that delimits not only to what extent reform can happen, but what kinds of arguments, and kinds of research, can effect change. Having such a yardstick against which to measure, and reject, other forms of understanding serves to counteract change of a more deep-seated nature. As such, the ways in which "real" science debunks critiques is nothing less than stultifying, because such a stance about what science is and who can rightfully do it and contribute to how it is constituted not only maintains it in its own image, but also deadens it to change and consigns it to a stasis that saps much of the creative impulse needed for technological innovation. And, as the case of engineering expertise suggests, when such a science becomes "the" science, practitioners whose scientific work will be well suited to that of many industry positions are marginalized by the very institution that claims for itself the right to anoint scientists. And, in the final analysis, those engineering educators with the least interest in listening and learning about the possibilities of reform of engineering education seemed the most likely to use blatantly subjective stances to defend the status quo, to argue from their own experience, instead of from their principles. But, such arguments nonetheless captured the imagination of others and held sway, and many were ultimately blinded to other possibilities.

All of this, then, comes full circle to suggest that such interpretations of my research missed an important finding that should have been comfortable even to positivists, because it has deep applicability and transferability (the next best thing to generalizability and universality). I found that different forms of knowledge coexist and are co-produced via social interactions imbued with relations of power in particular contexts, and one form of knowledge could be in ascendance in one context (academic-science engineering expertise at PES) whereas another might prevail elsewhere (design-engineering expertise in work settings). Thus, some are made "real" and others "strange," and this could be elbows over tea-kettles in another context, precisely the sort of commonality (general statement) that positivists profess to value, but in my experience one that many engineering educators have been unable to notice.

Notes

1 http://www.worldofquotes.com/author/Doris-Lessing/1/index.html
2 Insertions in quotes from the data set include explanatory text from the researcher, which is in square brackets, and descriptive text, which is in parentheses.
3 For instance, students could talk at length about the myriad ways that women had extra burdens compared to their men colleagues, but characterize this as equitable, saying, "We're all equal engineers, here."

References

Bucciarelli, L. L. (1994). *Designing engineers.* Cambridge, MA: MIT Press.

Howe, K. R. (1998). The interpretive turn and the new debate in education. *Educational Researcher, 27* (8), 13–21.

Howe, K., & Eisenhart, M. (1990). Standards for qualitative (and quantitative) research: A prolegomenon. *Educational Researcher, 19* (2), 2–9.

Law, J. (1987). Technology and heterogenous engineering: The case of Portuguese expansion. In W. E. Bijker, T. P. Hughes, & T. J. Pinch (Eds.), *Social construction of technological systems: New directions in the sociology and history of technology* (pp. 111–134). Cambridge, MA: MIT Press.

Lincoln, Y. S., & Guba, E. G. (1985) *Naturalistic inquiry.* Newbury Park, CA: Sage.

MacKenzie, D. (1996). *Knowing machine: Essays on technical change.* Cambridge, MA: MIT Press.

Shank, G., & Villella, O. (2004). Building on new foundations: Core principles and new directions for qualitative research. *Journal of Educational Research, 98,* 46–55.

Tonso, K. L. (2007). *On the outskirts of engineering: Learning identity, gender, and power via engineering practice.* Rotterdam: Sense.

11 Diversity of Knowledges and Contradictions

A Metalogue

Angela Calabrese Barton, Karen L. Tonso, and Wolff-Michael Roth

In the preceding three chapters, the authors presented ways in which science (and engineering) came to bear on their lives in three different settings: a mother concerned with the well-being of her baby; a person struck with a chronic (and for a long time) undiagnosed, disabling illness; and a long-time engineer turned educator in an ethnographic study of engineering education. Although some readers may come to this book thinking that such accounts are utterly singular, the very fact that the stories can be told (in language) and understood testifies to their cultural possibility and therefore to their wide applicability. Here the three authors reflect on the key issues arising from their chapters and the implications that arise from them for lay and professional educators concerned with science for all.

*

Michael: One of the key features that arises for me from the three chapters is that of emotion and orientation. What we have written about (and how we have written our accounts) is marked through and through by emotional qualities of life. And these emotional qualities of life—which traverse both knowing and being concerned with the things that (pre-)occupy us—provide particular orientations toward life. These emotional qualities, however, are totally lacking from the way science curriculum is thought of, designed, and enacted in today's classrooms.

Karen: Since the title guides us toward issues of identity and personhood, my remarks throughout will tend that way, and I will in particular add to this discussion a conversation about gender production and interpersonal relations, especially powered relations, because I think that this is a key social production of sites of science practice—a characteristic that our chapter situations share.

Like Michael, I was struck throughout our pieces by the pushing out of so much of what we might think of as authentic—that is, not idealized—personas. I read this as an indication of the many ways that the emotional, the personal, the contextual, the contingent have all been stripped via the application of the positivist paradigm—precisely because they cannot be objectified, universalized, and generalized. Objectivity, generality, and universality remain the tenets to which science of the positivist sort holds—what I shall call "the" science, here. That is, through sociocultural-historical processes true to "the" science, emotional orientations became irrelevant to the science project. It is worth noting that, at least in Western civilization, notions like emotion and personal orientation are associated with the feminine, that is, being emotional is for girls and to display it is to be "deficiently" masculine. So stripping science of its subjective aspects moves science into the masculine realm, or in other words, emotional orientations embody the antithesis of masculinity, at least the preferred (locally hegemonic) form of masculinity that holds sway in scientific fields. Here, it seems to me, it is important to be careful not to interpret femininity and masculinity as correlative with female/male (respectively), but to link them to the social construction of gender in and of science—how one makes sense of who acts the appropriate "woman" and who the "man"—what is appropriate to forms of gender in a field of endeavor or community of practice—and how one takes for granted that certain ways of "doing" life, particularly "the" science, become (are maintained as) masculine. Ultimately, such an analysis begins to unpack why "the" science seems not at all a "science for all people," but science for an elite of "the" science's production.

Angie: The entangling of position, identity, and knowledge construction that you reference here, Karen, challenges each of us to further unpack the multiple texts that make up each of our chapters. In my own story I described the value of standpoint theory in making sense of how I, as mother and science educator, drew upon mothers' collective knowledge to gain information and make important decisions regarding my daughter's health. To push this further, mothers' collective knowledge provided me with the resources I needed to stake some claim as an epistemic authority within the medical world. Despite a PhD in science education, at the pediatrician's office I am a mother. A woman. The patient (or the patient's legal guardian). The one who is told what to do. The strength of a collective history and knowledge offered me one set of tools to undo the "stripping of science of its subjective aspects" allowing precisely the subjective to direct medical next steps.

Karen: But, one of the things that was particularly noticeable in my research was the extent to which one persona tended to hold sway on campus—and this one would be quite inadequate for work in industry, ostensibly the ultimate clientele of the college, while another person would hold sway in industry—being formed underground, if you will, on campus. Thus, what is *made* elite is profoundly contextual. The preeminent engineer form differs in academia and industry. Industry calls for its preferred form, but engineering education produces its own.

Angie: It is worth stopping here for a moment to consider this point. The dichotomy set up between "school" and "practice"—or "academic" and "industry" works precisely to limit identity possibilities for it denies the possibility or the desirability of a hybrid identity. At the same time it works to further cement the genderization of engineering. Building an engineering presence through industrial experience requires at least implicitly access to the field—the cultural practices, language and otherwise, of practicing engineers. Who has this access/knowledge? And why is it that the "great equalizer" is downplayed in its value in the construction of an identity? These are questions I want to ask of the engineers in your study.

Karen: Furthermore, that these aspects of, in my case, industry practice and forms of being in engineering sites of practice seem, as Michael suggests, totally lacking in the way science curricula are thought of, designed, and enacted in today's classroom is true enough, but there is a rhetoric[1] to the contrary that I hear in many conversations with good science educators. Here, the constructivist urge acts as if it meant to include authentic persons—especially to bring the personal into science—but ultimately only included students' perspectives, prior experiences, and other sociocultural knowledge as a way to connect students to and lead them toward "the" science, never intending to change "the" science. Thus, the rhetoric attending constructivism, and possibly experiential science, serves eventually to maintain "the" science. And of course, this implies that it never intended to suggest that "the" science might be an inadequate representation of the world for literate activity in it. I wonder if, ultimately, science education/educators acted as if an awareness of the myriad ways that an individual's experiences of science might inform their actions in everyday life were important only to the extent they could be exploited for the service of "the" science, and otherwise were trivial. This implies two things, I think. First, it seems that authentic persons don't really matter. And, second and more importantly for these three chapters: "the" science—with its push toward a watered down and simplified reality and its addiction to rule-bound idealisms and idealized personas consonant with its own images, or (in the absence of rule-bound idealisms) its use of a statistical norm, which assumes that out of wide individual variation a representative mean will stand in for every individual—simply failed to prove useful in the fullest interpretation of scientific literacy, especially a literacy for all.

Michael: One of the key concerns I have with science education is that it is little different from what religious education has been in the past. It is a form of indoctrination into a form of thinking about the world that has been shown to be detrimental to the well-being of our planet that we both constitute and that is our home, our dwelling. In Greek, home and dwelling was denoted by the term *oikos*, which is the root of the term eco-; and knowledge (*logos*) about our dwelling and its inhabitants ought to be a central concern for "science." However, this ecological approach and ecological knowing was precisely absent from some of the discourses in which the three of us partook or in which we were the topical subjects. Angie's pediatrician did not consider the system Angie-Frankie-home-Internet group much in the same way that my own physician and specialists each looked at only one variable at a time, and they did so in only one abstracted part of the system as a whole, my body. Similarly, Karen's engineering professors did not look at the life and experiences of their students, much less at the engineering content of the papers that the groups handed in, instead focusing on superficial aspects of student work, such as writing mechanics, and from this, assigning grades and looking at their students in the form of these grades. I learned from these three chapters this: curriculum designers and policy-makers need to be more concerned with science as a contested terrain rather than with science as an uncontested and value-free tool.

Karen: Well, of course, I agree here in the main. However, it is important that we be careful what we mean by religious education, because not all religions are so narrow as Catholicism and other orthodox religions. I think what you mean to highlight might be better termed "orthodoxy." And yes, "the" science is orthodox and anyone who holds to other views can readily be made into a heretic. But it seems to me that the push toward universalizability and generalizability blinds purveyors of "the" science to the world's reality. When scientists fail to see the real world, they tend to see the one that they can mathematize and always get the "right" answer. Interest in messy worlds disappears, because that would be, well, too messy, and how in the world would one teach that?! (An argument I've heard often.)

Angie: I am often confronted with questions such as, "Well you cannot deny the fact that gravity is gravity. You can't contest that. If you leap from a cliff you will fall down." True enough but that misses the point about a contested terrain. Ever since my encounter with the first group of youth I taught in homeless shelters, I have found much value in the idea of science as contested terrain. It captures the competing discourses alongside the complex ways in which knowledge is situated and produced. To return to the point about gravity, speaking of science as contested doesn't change the physical phenomenon of falling off of a cliff. It does however open up how or why we use our understandings of gravity to talk, think, and do certain things. Take cluster bombs for example. I pick that because the US does not want other countries to have these, but the country is not willing to give up their own. A simple Wikipedia definition tells us that "Cluster munitions or cluster bombs are air-dropped or ground-launched munitions that eject a number of smaller submunitions ('bomblets'). The most common types are intended to kill enemy personnel and destroy vehicles. Submunition-based weapons designed to destroy runways, electric power transmission lines, deliver chemical or biological weapons, or to scatter land mines have also been produced. Some submunition-based weapons can disperse non-munition payloads, such as leaflets." My point here is that the knowledge we construct around gravity takes many forms, with many implications. But this is a simple example. What about the knowledge we construct around what healthy levels of tropospheric ozone are? Who set these limits and how were these limits set by the interpretation of data sets? Likewise, who has the power to label a site a superfund clean-up site?[2] How much pollution and illness does it take to get labeled a superfund clean-up site? I could go on, but the point is it is not enough just to know what gravity is, what ozone is, or what a superfund clean-up site is. It is how these ideals are constructed and used historically that give shape and meaning—and access—to discourse and power of science.

Karen: Yes, indeed! For instance, there is a well-known use of gravity—an equation—to characterize how an object of known mass will move in a particular gravity field. But, this equation neglects the effects of the air through which the object falls. In this idealized situation, a man with a parachute (mass *M*) falls at the same speed as a cluster bomb with the same mass, which of course is not the case—if the parachute opens. Air matters in the world where we live and have our experiences.

This is what I mean when I say that positivists produce reductionist, essentialized renditions of the world. But I think that the real danger from this over-simplification is that students of science may cease to think with a curious mind—what we, as science teachers, ought to nurture if we hope to create the very best kind of scientists. And, all too often in the engineering senior design class, this was the case for students whose primary attention was on their academic-science classes: they had lost interest in the world of engineering projects and worried almost exclusively about the world of conventional coursework and tests, and mathematized versions of the world. And, ultimately, they *became* certain sorts of engineers.

In the case of medicine, I wondered if Michael and Angie weren't captured by the business of medicine, which (we are told) cannot succeed with a practitioner who spends too much time with patients, who sees the world in complex ways, who allows others to act as if they are as knowledgeable. Only the simplification allows the "proper" progress of the medicos, here, it seems.

But, not all doctors or science educators are of this sort, and I can think of one quite different case. And his medical preparation included a long stint in the Public Health Service on the Navajo Reservation, where he learned quite a bit about alternatives to conventional medicine. When I presented him with a disease much like Michael's (after having developed a deep and trusting relationship around the care of my five children), Dr. Mac listened to me and we collaborated and co-produced my care, diametrically opposite to Michael's experience. But, the point I think we are making is that doctors of any other sort tend to be thought of, in some ways, as improper doctors, which is the point of Sherryl Kleinman's (1996) work in an alternative health organization. Thus, a cultural production theorist (such as myself) recognizes that people not only have their own interpretation of the world, which becomes evident in their actions, but also it/they can be made into an inadequate set of understandings or ways of being by cultural, historical, social, and political contexts with entrenched meaning-making processes.

Angie: Standpoint theory teaches us to pay attention to how understandings and explanations of the world are always situated and written from the "standpoint" of the social agent. I recently read that Howard Dean, the Chairman of the Democratic National Committee in the US, has had trouble playing the role of "conciliator" in the party primarily because he is a doctor. Doctors are taught to tell people what is wrong with them and how to live their lives. They are not taught to listen.

Michael: These are precisely the messages I recently got from two very different directions. On the one hand, my wife is talking about some book she is reading at the moment written by a doctor about doctors and how these colleagues fail to listen. On the other hand, an old friend who married a doctor says precisely this, that her husband does not listen, he is trained to tell people what to do.

Angie: This is why the story of Dr. Mac speaks to the value of learning to listen as well as to the value of learning to co-construct new medical narratives based on individuals' stories. Learning to listen and building new knowledge from such stories, though, is a main reason why standpoint theory is so controversial. Standpoint theory espouses that politics are an important element of knowledge production, and that the concerns of oppressed groups are not just social and political in nature but intimately linked to knowledge production itself and its relationship to power. Just think—your story or the stories of the Navajo might actually bump up against medically established norms. Historically, people have been told to re-read and re-write their own experiences. Here take this medicine. Yet, the controversial nature of standpoint projects is what gives them value and power: science is contested and norms are never outside of context. Listening to stories like yours should give us pause in any of our efforts to norm, whether it be in medicine or education.

Karen: On another front, it seems that one of the beauties of Angie's group was that it was large enough and had varied enough experience to have two members whose children had the same condition as Frankie—a statistical power of a sort, large enough to encompass a particular medical diagnosis. Thus, unlike doctors who see a wide range of situations and work from the mean, a focused mother with only a limited number of children to follow held a very detailed set of understandings about her particular realm and collectively this group could be a boon to a parent whose child's doctor wasn't listening. Learning to assert one's expertise, as a mother, is a valuable skill, as is cultivating chutzpah—but these are hardly qualities that endear science students to their teachers, I suspect. Thus, it is important to be able to remind doctors (and science educators) of their fallibility and their lack of contingent, contextualized knowledge and to assert one's own, which is a larger lesson learned from Angie's experience that can be taken up in science classes. Imagine if one acted as if kids knew things! Rather than always feeling like they weren't living up to the standardized-test expectations, based on some norm that may or may not be appropriate, kids would feel like they were somebody who knew stuff. Wouldn't classrooms be more humane places if this were the case, if all kids came to realize that every other kid knew something?

Michael: Just as you speak about asserting one's expertise, I am wondering how we can come to "integrate" different knowledges in ways so that one form of knowledge does not win out over others, or one is changed and made to fit another form when they are contradictory—which is what some approaches to knowledge integration do when they take into account traditional ecological knowledge if it can be expressed in scientific terms. Truly democratic forms of knowledge integration emerge when we learn to take into account knowledges that are contradictory and each as a one-sided expression of the system under consideration. As a science teacher, I have had epistemological discussions with my students even though, in the end, it was the grades that mattered to them, their parents, and the school.

Karen: Also, I chuckled about that "oh, you poor benighted mother" talk that the doc gave Angie about holding the baby's head. I received one of those when my oldest daughter was three months. I said: "I'm afraid she's going to hurt herself when she climbs out of her crib." (She was my first, so I didn't know that few babies climb out of cribs at three months!) The doctor gave me the "just keep her head in the crook of your arm so that her head is protected" sermonette. I replied: "I don't think you understand; when she wakes up from her nap, she climbs out of the crib and drops to the floor." Had a chair not been there, he'd have sat on the floor. He was quite adrift, having little experience with the art of living with children who are outliers to the mean. I suggested taking the crib out of the room, putting the mattress on the floor, and putting a screen door on the room. She'd have been over a gate in a couple of days, his suggestion. Like Angie, Michael, and our science and engineering students, I knew things that he didn't and he certainly failed to accord respect for my understandings—the very core demand of democratic citizenship and multicultural awareness.

I didn't like the implication that I was an idiot who didn't know how to hold a baby, and if they were really thinking I was a moron, I found it incredible that they'd sent this precious baby home with me three months before. This situation paralleled my student-engineer participants who used to turn to one another, hold the hair off their forehead, and say to a classmate, "Do I have moron stamped up here?" Bottom line, it doesn't matter whether you have a Masters in physics, are a respected science educator, an engineer, or just plain folks; doctors, other scientists, and science educators shouldn't treat folks like morons. We know stuff and it adds insult to injury when they act as if they are the Great Oz. We saw the movie and have them figured out. Wouldn't sites of science practice—medical offices, classrooms, engineering campuses, and so forth—be better places if practitioners gave up the pretense of omnipotence emerging from productions of social power in science?

Angie: Karen and Michael, you both challenge us to consider what would classrooms be like if kids were positioned as knowledgeable, or what would doctor visits look like if patients were treated as knowledgeable. Underpinning this powerful question are the dichotomies raised by each one of our chapters: Engineer or student? Patient or doctor? Unknowledged or knowledgeable? Unskilled or skilled? These dichotomies work to refuse efforts to build hybrid spaces that allow for and necessitate the blending of these stances. Further, the very act of breaking down these dichotomies—of creating hybrid identities that allow for multiple representations to coexist—are always political and of the highest risk for those whose knowledge, discourse, and identities are positioned as lesser.

Michael: Coexistence is what I was aiming at in my earlier comment and the possibility of living with and accepting contradiction. What we need to learn is how to make decisions that take into account this hybridity of knowledges, for we tend to base them on an "I-do-X-because-Y" structure. If there are two contradictory forms of knowledge Y—i.e., Y_1 and Y_2—suggesting different forms of action—i.e., X_1 and X_2—there is a conflict.

Angie: Individuals in any given community of practice engage in acts of hybridity all of the time when confronted with differences drawing upon multiple resources or funds to make sense of the world, whether or not their actions are actually recognized as such. Yet, being "in-between" several different funds of knowledge and discourses can be either productive or constraining, and even marginalizing, depending upon how they are recognized by those in power (Moje, Ciechanowski, Kramer, Ellis, Carrillo, & Collazo, 2004).

Michael: In my own experience with the two medical systems, the regular and the alternative one, I have felt the effects that the "scientific," single-variable approach to thinking about problems. My experience made me think about other problems that single-variable approaches of science have caused in this world. Thus, the cane toad was introduced to Queensland, Australia, in an attempt to combat or eradicate the cane beetle. Because of its extremely poisonous skin, however, the cane toad does not have a natural predator and, in spreading all over Australia, has become a major ecological disaster. There are many other examples of ecological destruction caused by humans generally, and by science and scientists in particular.

Karen: Yes, yes, I definitely agree—the saltcedar (Tamarisk), the gypsy moth, the starling, but allow me to take a slightly different tack here, one related to where you started. What I want to highlight is how difficult it is for someone trained in "the" science to act otherwise, to eschew it in favor of other ways of scientific literacy. In particular, I wanted to note the difficulty with which we give up this single-variable approach, even when we ourselves recognize its limitations. Notice Michael's quandary, for instance. On the one hand, he is most upset that his various medicos cannot take *him* into account—seeing only the aspects of him that their "polarizing lenses" make visible. From direct experience, I have enormous empathy for being the elephant that the seven blind men are trying to identify. On the other hand, notice that when he wanted to sort out if some food or another was giving him trouble that he resorted to a one-variable approach. I do not mean to imply that Michael is inconsistent etc., but to note that there is every likelihood that this either/or stance—*either* science with its parsing into bits and simplification of complex systems; *or* a holistic approach trying to account for almost everything simultaneously—may be at fault. Instead, increasingly I wonder if both are not necessary, an issue I would like to return to in a moment.

Angie: Situating individual variables within a larger, complex system may allow for multiple pathways to intersect with single variable allowing multiple story lines to emerge. I often wonder if the challenge we face on the frontiers of science is less about the variable and more about the (singular) story.

Michael: I also feel that the commercial and ethical aspects of science are often overlooked, and that students ought to know more about these aspects of science that are so highly value laden and of an ethico-moral nature. As a critical educator, it would be important for me to bring out science as an epistemology that ought to be questioned—as I have done with my students as a high school physics teacher—and that leads to a much more critical citizenry when it comes to various forms of knowledge: religious, technological, legal, ethico-moral, political, environmental, and so on. I would like my "science" students to be able to identify environmental polluters, and to make knowledge about the degree of pollution and the identity of the perpetrators known to the community as a whole. We can *educate* (rather than *school*) our students to have the same kind of emotional volitional orientation to their community and its environment as the three of us have shown towards issues that concern our selves, our health, and our knowing about issues of concern.

Karen: Having training as a political philosopher of education (a subfield of ethics) and knowing something about philosophy of science, I question whether these are simply epistemology issues. I implicate them as metaphysical ones, with ethical overtones. Again, I think that it's the either/or choice being forced between science and holism that is deadly. I maintain that we need to know how to integrate the pretense of a rational, simplified science *and* the mucking about in the mess of a chaotic, irrational world. This is the message of the design-engineering practitioners. The very best students in design projects, those I termed "design engineers," held an enormous store of formal knowledge of the sort "the" science values to the exclusion of other forms. But, and this is the key, these engineering practitioners *also* held a wide range of other forms of knowledge that they acquired from people in the field, from earlier experiences, and from understandings of equipment and instruments. I think that this is the dilemma that the feminist critique of science never quite got right—that we must choose between science (which has historically been associated with masculinity) and alchemy (which is holistic, particular, contextual, contingent, and associated with the feminine).

Michael: Let me stop you here for a second. This is the problem when you accept multiple competing or contradictory forms of knowledge. When you have to make a decision, then you appear to be forced to go one or the other way. The thought of Schrödinger's cat comes to my mind, we can accept it as living or non-living, a mix of states well described by the quantum equations. But as soon as we look, when we make the decision to look and actually do it, then the mixed state collapses—in physics talk, this is the collapse of the wave function—and then the cat is either dead or alive.

Karen: Let me return to the feminine. Muriel Lederman and Ingrid Bartsch (2001) summarized how an androcentric science came to be created. The gist of the feminist analysis holds that before the Scientific Revolution what counted as science was an endeavor closely linked to the real world, in all its complexity, where one asked "why" questions as an embedded observer. Time was seen as cyclical, and important issues were particular and contextual. Overall there was a valuing of both prototypically masculine and feminine virtues, a Hermaphroditic take on the world. Compared to this, in the new science that emerged post-Scientific Revolution, a detached observer asked "how" questions, time was seen as linear, knowledge worth having was general, and the virtues that underpinned such an endeavor were those held to be prototypically masculine, especially hierarchical, that is imbued with enormous inequity, in fact designed to be unequal (hence the ethical concern). David Noble (1992) linked the form of academia associated with "the" science with monastic life, which further intensified the acceptance of certain forms of men's lives (aesthetes), as well as exemplified hierarchical social relations and distancing scientists from the realm of "ordinary folks." To my way of thinking, this is about metaphysics, assumptions (axioms) about the nature of a world worth knowing.

Thus, as it came into being in its modern guise, "the" science pulled away from the particular and the personal (the so-called "private" sphere, typically associated with the feminine), and seemed to distance itself from the political and the ethical dimensions of life. On this we agree, I think. But I wanted to highlight that the side that you hoped would engage your students—environmental activism, for instance—has long been associated with tending toward femininity, at least affiliated less and less with a militaristic "lord of all he surveys," or an Old Testament (Genesis) god-like visage: "Let us make man in our image, after our likeness: and let them have dominion over the fish of the sea, and over the fowl of the air, and over the cattle, and over all the earth, and over every creeping thing that creepeth upon the earth" (Genesis 1: 26). This "lordly" image of masculinity seems precisely the one associated with colonization (and the theft of mineral deposits and other raw materials from undeveloped countries, and the inattention to the downside of mine dumps, etc.) and leads to the present-day paternalism embedded in engineers' deciding what dams to build and where, in doctors' determining what diseases to study and how, and so on. This is a manifest destiny sort of science, not a science for all; this is a science working in the trenches for the mighty, a science that lines the pockets of those in power and able to take from others.

Critical theorists have long argued that appeasing words, such as calls for a science for all, are a rhetoric that diverts attention from the reality that when you become one of the members of the science club, there is every possibility of having one's otherness co-opted. In fact, many critical activists from under-represented communities speak about having to give up who they are to gain membership. In fact, in the liberal philosophical framework, there is a tendency to expect that when the mix of insiders changes there will be a change in the way that science is done. Yoder (1991) found precisely the opposite. Rather than women's acceptance increasing as they demonstrated competence and their numbers grew, backlash in male-identified work and learning settings became more blatant and overt discrimination more intense, which Clara Bingham and Laura Gansler's (2002) description of Lois Jenson's experience in Minnesota mines exemplified. Thus, rather than "the" science becoming a place where "others" can contribute to how science is constituted, "the" science is remarkably obdurate in many sites of practice, as evidenced by our writings here. And, this circumstance highlights the impossibility of dismantling the master's house with the master's tools, because one must overthrow the axioms of objectivity, generalizability, and universalizability, which simply are not on the table for discussion.

Angie: It is interesting to me here that my first reaction is "I'm not a political philosopher, so what would I contribute to our conversation at this juncture?" Yet as I think this thought I laugh because it cuts to the core of the issues we are bandying around. What does it mean to be an epistemological expert? What does it mean to learn to take a stance? I am compelled to pull us back into the realm of school science because there is such a dominant discourse in the US (and arguably the developed world) that in classrooms there is a tradeoff between allowing students to have agency and maintaining standards. In other words, if you allow students to take action with/in science then somehow you are skirting the science standards. How do we implode this dichotomy? Michael you state that: "As a critical educator, it would be important for me to bring out science as an epistemology that ought to be questioned . . . I would like my 'science' students to be able to identify environmental polluters to make knowledge about the degree of pollution and the identity of the perpetrators known to the community as a whole." This stance transcends the dichotomy situating knowledge production as a critical component of taking action. Yet, Karen, you remind us that this process is also about taking an ontological stance. Again, this process is often the most risky for those who have been most marginalized. This latter dichotomy is one we ought to return to, expose, and open up for consideration.

Michael: The call for science education to serve all students (e.g., "Science For *All* Americans") to me fails precisely because it does not take into account personal experiences, the lives, works, and voices of individuals. We know from the research on situated cognition that our orientations and concerns mediate our interest in expanding our room to maneuver; and this expansion in room to maneuver constitutes learning. Without the emotional-volitional and ethico-moral orientation that we take with respect to the real issues and concerns of our lives, knowledge is inconsequential and therefore without value. It is more important to know how to access disembodied knowledge, which because of its very nature can be stored anywhere—e.g., the Internet as my case study shows—rather than in the mind and body of the person. There is just too much factual knowledge out there relevant only in very particular context to know even a tiny fraction of it. Knowing how to access and critically interrogate are more desirable curricular foci than knowing any (in all but a few instances irrelevant) factual knowledge.

Angie: I think what is missing here is knowing what debates are worth having. Take for example, the case of global warming. What does one need to know or be able to do to recognize this as an important debate and to have the capacity to know how to seek out ideas, and so on for constructing a thoughtful and reasoned stance on global warming? Arguably one should know what a good question is. One should know what solid data look like. One should know how to interpret these data. I am not arguing that "what global warming is" should necessarily be taught in schools. However, I think we have to stand back from our points of privilege as science educators to ask what do we know that allows us to engage with some ease in debate on scientific issues? What science practices and or knowledge (or the experience of mastering some knowledge) helped position us this way? It does seem that I am invoking the master's tool here, of mastery and reasoning, but not without a critical awareness, or at least a critical mindset to questioning what makes it possible to accomplish. As Michael suggests, "Knowing how to access and critically interrogate are more desirable curriculum foci than knowing any (in all but a few instances irrelevant) factual knowledge."

Karen: I do not disagree with Michael's assessment of the reasons for the failure of science educators to serve all students, but argue that science education's cleaving to hierarchical relations thoroughly counters anything like an egalitarian outcome implied by the "for all" part of "Science For All." Clearly there was no egalitarian mission represented in my research findings about engineering education (nor in the interactions Michael and Angie had with their doctors). As carried out in the campus studied, engineering education was designed precisely to "weed out the weaklings," "separate the men from the boys," and otherwise produce an elite rank in society—the mystique of the geek. PES lauded its ability to create a particular form of hierarchical life that existed on campus, even as many there—students and faculty—could mouth a rhetoric of equality, saying for instance that "We're all equal engineers" or "We treat all students the same," while describing in great detail enormous gender inequality, among men and among women, as well as between women and men. Thus, PES engineering education strove not to recognize the variation and wide range of experiences of its students, but to serve as a normalizing technology that acted as if uniformity of students was not only the norm, but also the revered outcome. So, yes, science for all cannot succeed when so little respect is accorded authentic personas.

And this was demeaning to student engineers who found other ways to enact themselves as engineers, as was the way doctors treated Michael and Angie. Even when students could, as you put it, "access and critically interrogate" information, the campus could not see them, recognize the value of their particular knowledge, or understand that this was equally central to being an engineer when coupled with "academic-science" expertise. Neither form of knowledge was sufficient and only in tandem could students demonstrate engineering expertise of the sort desirable in industry settings. And, yes, the ability to ferret out information that is given in textbooks provided students with the information needed to complete their projects. This was the sort of life-long learning that engineers will need, along with the ability to follow technological innovation in their fields and to jump from field to field as economic shifts necessitate.

It seems to me that, as consumers of any number of scientifically grounded practices and goods (medicine being one of many examples, with petroleum products, cleaning agents, pharmaceuticals, and foodstuffs coming to mind), citizens must be able to follow advances in the field, to read websites such as drugline.com, and research illnesses on medline.com, just as scientists and engineers and science educators must be able to "read" the citizenry (via on-going social interactions where respect for others as equals *and* recognition of their unique understandings exist) and incorporate their diverse needs, interests, and desires. Ultimately, science and learning science is far more than what happens in most science classrooms, and more classrooms need to admit to that, begin to understand the limitations of "the" science toward which they aim, and in due course re-envision "the" science typically reserved for a scientist elite, and see it as a science for—and about—all.

Notes

1 And here, by "rhetoric," I imply a statement or process that, at least on the surface, seems to imply "A," while the reality of the situation overall tends toward "not-A." Thus, a rhetoric is a story that insiders tell to convince themselves that they are not doing what they are in fact doing.

2 Superfund is the federal government's program to clean up the nation's uncontrolled hazardous waste sites.

References

Bingham, C., & Gansler, L. L. (2002). *Class action: The story of Lois Jenson and the landmark case that changed sexual harassment law.* New York: Random House.

Genesis, Bible. Downloaded May 19, 2008 from http://scriptures.lds.org/en/search?search=dominion+over&do=Search.

Kleinman, S. (1996). *Opposing ambitions: Gender and identity in an alternative organization.* Chicago: University of Chicago Press.

Lederman, M., & Bartsch, I. (Eds.). (2001). *The gender and science reader.* New York: Routledge.

Moje, E. B., Ciechanowski, K. M., Kramer, K., Ellis, L., Carrillo, R., & Collazo, T. (2004). Working toward third space in content area literacy: An examination of everyday funds of knowledge and discourse. *Reading Research Quarterly, 39*, 38–70.

Noble, D. (1992). *A world without women: The Christian clerical culture of Western science.* New York: Knopf.

Yoder, J. D. (1991). Rethinking tokenism: Looking beyond numbers. *Gender and Society, 5*, 178–192.

12 Appreciating Difference in and for Itself

An Epilogue

Wolff-Michael Roth

Throughout this book, the authors argue not only for situating science and science education in the knowledgeabilities that students specifically, and ordinary people generally, develop throughout their lives but also, and more importantly so, they make a case for forms of science and science education that really matter in everyday life. As the contributions to Part I show, these experiences are inherently situated, both culturally and historically. This immediately leads to contradictions, because, as three decades of conceptions and conceptual change research have shown, ordinary people talk about natural phenomena in ways that are inconsistent and incompatible with science.

We can take a position according to which ordinary people just get it wrong and the task of science education is to fix them by "confronting" them with their misconceptions, to encourage them to abandon these in favor of the scientific way of knowing. But this position fails to recognize that a strong possibility for the continued appeal of non-scientific language for explaining natural phenomena lies in its usefulness for dealing with the everyday world that surrounds us.

A parallel with the relationship between high, literary language and the language of the common folk (Bakhtin/Vološinov, 1973) constitutes one of the ways in which we may productively rethink the relationship between different knowledges generally, and that between scientific and folk knowledges more specifically. We may take another hint from the relationship between the high literary genre of the novel and its refraction in the parodies that have accompanied one another for millennia (Bakhtin, 1981). In each case, literary language and genres relate dialectically to everyday language and genres, none expressing the phenomenon of language and genre in its entirety. Each term is the necessary complement to the other, each being only a one-sided and partial expression of the whole. Because of their continuous relation, the two sides work each other, rub up on each other, transform one another, one incorporating the other.

Difference in and for Itself

Throughout Part I of this book, we encounter different forms of knowing that have evolved in the course of the respective cultural histories: agricultural knowledges of Mexican immigrants (Richardson Bruna & Lewis) or Zuni farmers and Western gardeners (Brandt); the understandings of plants among African Americans in the deep South (Carlton Parsons) or the everyday understandings of society among African American city dwellers as expressed in hip hop (Emdin); and the cultural knowledges that new (Asian) immigrants to North America bring with them and the tensions that arise from the encounter with scientific language and culture and their everyday equivalents that are used in teaching (Hwang/Roth). To date, many science educators and science teachers have not yet come to the point of being able to appreciate the differences within and between these cultural forms of knowing and talking about the natural world. To truly appreciate the differences, difference itself needs to be properly theorized: as something that exists in and for itself.

In Western epistemology, (self) identity is the fundamental starting point of theorizing. That is, the starting point for thinking is the identity $A = A$, which identifies what is different as *not-A* (or $\neg A$). The two resulting terms, A and $\neg A$ are related by an inequality: $A \neq \neg A$. A third possibility does not exist (the Romans used to say, "tertium non datur," a third option is not given). Here, then, difference is defined in terms of different ("≠") from identity. Difference does not exist in and for itself: it is but the inversion of identity. Difference is everything that does not fall within identity. Difference thereby is not appreciated for what it is, independent of, and perhaps prior to self-identity. Applied to science education, this means that we cannot appreciate what is different from scientific discourse or scientific conceptions. It is not surprising, therefore, that the science education discourse uses the same logical feature to characterize student understanding prior to (and often even following) instruction: with respect to a scientific conception A, the expressions *mis*-conception or *alternative* conception point us to what is *not* ($\neg A$). Here, what is not is real science knowledge or real science discourse. Other discourses and cultures are not appreciated for what they are, in and for themselves, but are acknowledged only as something different to be restructured and overcome.

Philosophers in the 20th century have come to realize that the idea of self-identical entities is not very fruitful for understanding cultural phenomena generally and linguistic phenomena more specifically. Thus, there cannot be something like *a* (self-identical) culture or *a* (self-identical) language. In fact, our everyday experience with language shows that it is not self-identical. For example, when a student asks her science teacher, "What do you mean by saying X?" she requests a translation *within* her native tongue such that she can understand what the teacher means once she hears his second (third or nth) expression. Translation implies difference, but this is a translation *within* language, pointing us to the inner difference of language within itself. *Two* (different) expressions can be used to mean the *same*, but the student does not understand the first one but does understand the second (third, nth).

In the same way, the same expression (word, utterance) may be used to mean something different. In fact, the repetition of a word always is associated with a different meaning (Bakhtin/Vološinov, 1973) precisely because of the repetition: the repeated word is heard with respect to its earlier occurrence, which constitutes a context, whereas in its first occurrence, the word does not have itself as the context. In the second occurrence, the word comes to be inflected, refracting its own earlier existence, changing at the very moment that it is repeated. This is precisely the message of *Difference and repetition* (Deleuze, 1968/1994), a work that theorizes difference in and for itself.

We are now precisely at the point where we can appreciate the quote that I used to open Part I: "Every culture is in itself 'multicultural,' not simply because there always has been an antecedent acculturation but more profoundly because the gesture of culture itself is a gesture of the mêlée" (Nancy, 1993, p. 13, my translation). The mêlée comes from the fact that culture is not self-identical but internally contradictory, its mutually constitutive parts in continuous exchange, changing one another.

Implications for Teaching Science

One of the most frequently asked questions may be the one about the implications of research for teaching. A not so well-meaning reviewer of the proposal for this book suggested that the contributors were writing for themselves, making important points based on significant research findings but were "not truly impacting teachers' practice and the significant classroom events that define contemporary science education." This quote contains the whole crux that we, the contributors to the book as a collective, attempt to address. The point of a science education *from* people *for* people is not to "impact"—read "cause to change"—teaching and learning from the outside, but to develop a way of understanding the contradictory nature of the task of teaching science, all the while acknowledging incommensurable prior and alternative experiences and discourses as the very necessary ground of any learning.

There are many science educators and science teachers who acknowledge the need to ground teaching in the lived experiences of the students. But they do not generally provide us with a description of what it might look like to *explicitly* draw on the experiences and existing discourses, which are incommensurable with science and scientific language, and *build on* these to evolve forms of understandings and discourses that are commensurable with current science. The traditional answer has been to organize tasks that challenge students' pre-instructional understandings and to lead to restructuring and conceptual change. The discourse of challenging and restructuring understandings already points to the potential real or symbolic violence involved in the science curriculum, which values the articulation of lived experiences and discourses only to overthrow them, so that students construct, appropriate, or information-process the correct ("authentic") science. Moreover, traditional approaches to science education reform do not thematize why students might take up the object/motive of learning science when the need for doing so is not apparent. These, however, are precisely the issues that the contributors to the present volume articulate and make salient. What can science teachers and educators learn from this inner contradiction? How can they address this inner contradiction in redesigning the learning environment that they provide for their students? Here, Mikhail Bakhtin may have some interesting answers.

Among other things, Bakhtin was interested in understanding and theorizing the change of one form of language to another, from one literary epoch to another. He was not interested in developing a model in which the change came from without language. Rather, he was interested in developing a model—in contrast to linguistics, which is concerned with structures rather than dynamics—where change is inherent. The model he and his associates in the Bakhtin circle proposed (e.g., Bakhtin/Vološinov, 1973) postulates changes in literary language to occur not from one novel to the next, one stylistic form (genre) to the next, but in each speech act ("utterance") of the culture as a whole. Each utterance—a novel or poem constituting but one of these—changes the language at the very instant of reproducing it. Rather than changing through a process of restructuring—in the way that theorists of conceptions and conceptual change postulate that learning occurs—Bakhtin describes change as continuous, repetition and difference in and of language being but two sides of the same coin. The changes in the literary languages of the high culture come to be but a refraction of linguistic change in a culture as a whole.

In a similar way, Bakhtin (1981) theorizes the changes in literary forms. These changes are continuous, arising from the dialectic relation between high literary forms and their concurrent inversions in the parodies of popular culture. The different literary forms that can be observed at the cultural-historical level are but static images of the continual changes that a language and the genres (forms) it enables undergo. We can use these ideas to inform a project of science (education) from people for people by people.

Let us begin by identifying the static images of beginning and endpoint: everyday knowing and experience, on the one hand, and scientific knowing and perception, on the other. If we take the corresponding languages (discourses) as genres in the way Bakhtin conceives them, then the change from one language (discourse) to another can be understood as arising from the continual changes of language at the very moment that an utterance occurs. But what we then observe are not pure forms—if we take each as a pure form rather than as inherently hybrid—but hybrid languages that no longer are those of the everyday and not yet those that are spoken and written in science laboratories and journals.

Once I had realized that learning was a process of continuous change, my actions as a high school physics teacher began to change. I no longer expected students to express themselves consistent with the domain after working through a single unit. Rather, my vision for outcomes became long-range. I told students, for example, that I expected them to be comfortable with and competent in doing and talking physics in the course of a year or two, at a point when I anticipated them to master the subject matter outlined in the provincial curriculum guidelines and statements of learning outcomes. That is, even if our—the teacher's—main concern lies with achievement on standardized examinations at the end of a course, this does not mean that students have to master each of its topics at the moment that it is the goal of the ongoing tasks. As a physics teacher, thinking in this way allowed me not to worry about the hybrid forms of talk students were using after a month or two in my physics courses because I knew that they were on a developmental trajectory. I also knew that, prior to the formal examinations, I would have preparation time during which students could consolidate the changes they had undergone for much of the school year. I observed what I initially called (following the philosopher Richard Rorty) "the muddle" and what I later denoted by *Sabir*, the continually self-hybridizing hybrid languages of people in the process of doing their business, whether this was that of the ancient Mediterranean merchants or those of my students learning to converse about physical or epistemological topics.

A concomitant point that needs to be observed is that of the object/motive. We can learn from the chapters in this book that people will learn when they know that what they do will expand their room to maneuver, that is, their agency. The frequently heard student question "What is this good for?" or "What do I need this for?" is an expression of the problematic here. Thus, Angie Barton or I did not ask "What is this science good for?" but we found that knowing more about *torticollis* or the role of vitamin B12 in the human body would expand our power to act, and our endeavor in learning more was inherently motivated in the process. In Carol Brandt's situation, it turned out that asking Anselmo, who took her down to the arroyo and told her to get some of this "tree-sand," was a much better approach than the one she could have gotten from a scientific approach to agriculture for producing a lush vegetable garden in the arid conditions of the Zuni lands.

Coda

Everyday understandings and discourses are terrains and resources upon and with which new entirely different forms of understandings and discourses can be built. This is precisely what happened culturally and historically when science emerged and developed from what subsequently has become denoted as non-scientific during the times of the ancient Greeks and Romans. These everyday understandings and discourses are to be appreciated for what they are, in all their difference from science in its canonical forms. These understandings and discourses have proven to be powerful, so powerful that they have not been abandoned—even though, like the Aristotelian discourse about momentum, physicists abandoned it centuries ago. Yet even physicists continue to reproduce non-scientific discourses, such as when they marvel at the beauty of a sun*rise* or a sun*set*. The fact that the scientific ways sometimes are not taken up into other cultures may indicate to us their limited usefulness or beauty. Why would everyday folk not pick up on science if they knew that it would increase their control over the (natural, societal) environment and over their life conditions?

Appreciation of between-culture differences means that we accept otherness rather than, with colonial zeal, attempt to eradicate the differences under the banner of science. There are lessons to be learned from Zuni gardening, hip hop music-making, Mexican agricultural knowledge, traditional ecological knowledge of African Americans, or the knowledges Asian immigrants bring with them to their new homelands. It is precisely out of the difference that continual change can arise, not only of everyday cultures, languages, and genres but the continual change of scientific cultures, languages, and genres as well.

References

Bakhtin, M. (1981). *The dialogic imagination.* Austin: University of Texas.

Bakhtin, M. M./Vološinov, V. N. (1973). *Marxism and the philosophy of language.* Cambridge, MA: Harvard University Press.

Deleuze, G. (1994). *Difference and repetition* (P. Patton, Trans.). New York: Columbia University Press. (First published in 1968.)

Nancy, J.-L. (1993). L'éloge de la mêlée. [Eulogy to the mêlée.] *Transeuropéennes, 1,* 8–18.

Index